30-SECOND
BIOCHEMISTRY

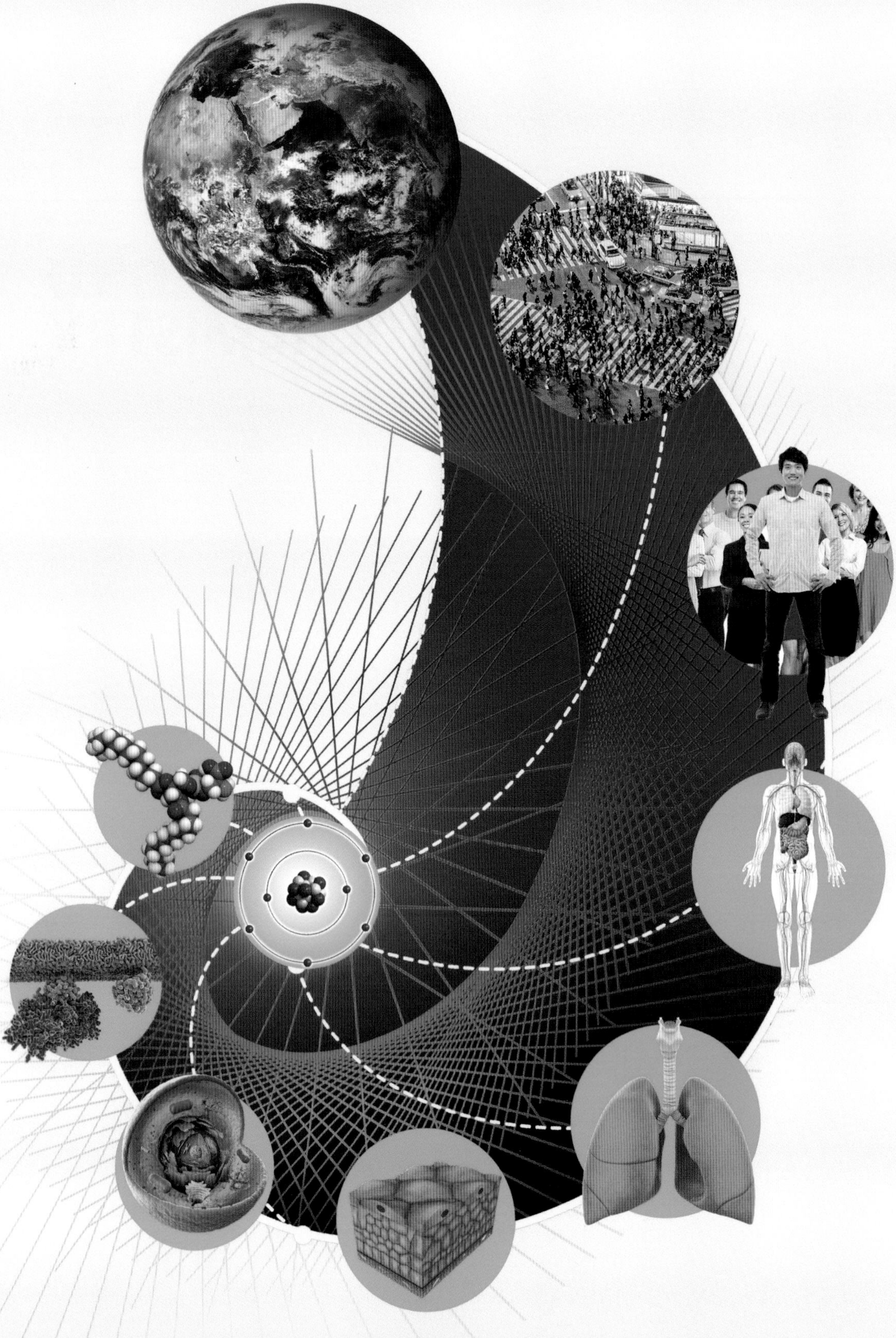

30-SECOND
BIOCHEMISTRY

The 50 vital processes in and
around living organisms, each
explained in half a minute

Editor
Stephen Contakes

Contributors
Stephen Contakes
Steve Julio
Kristi Lazar Cantrell
Ben McFarland
Vassilis Mougios
Anatoli Petridou
Andrew K. Udit

Illustrator
Steve Rawlings

IVY PRESS

First published in the UK in 2021 by
Ivy Press
An imprint of The Quarto Group
The Old Brewery, 6 Blundell Street
London N7 9BH, United Kingdom
T (0)20 7700 6700
www.QuartoKnows.com

British Library Cataloguing-in-Publication Data
A catalogue record for this book is available from the British Library.

ISBN: 978-0-7112-6367-3
eISBN: 978-0-7112-6369-7

Commissioning Editor **Caroline Earle**
Designer **Ginny Zeal**
Illustrator **Steve Rawlings**
Glossaries Text **Stephen Contakes**

Printed in China

10 9 8 7 6 5 4 3 2 1

MIX
Paper from responsible sources
FSC® C016973

CONTENTS

INTRODUCTION
Stephen Contakes

Biochemistry deals with life, a concept difficult to pin down but which may be roughly understood as comprising parts of the natural world that are distinct from their surroundings, can take in and use energy to sustain themselves, and reproduce. It examines life in terms of atoms and collections of atoms called molecules. Over the past two centuries the merger of chemistry and biology in biochemistry has yielded an increasingly detailed understanding of how living things work. This has especially been the case from the twentieth century onwards as chemists and physicists have developed powerful tools for probing the composition and structure of matter at the molecular level. Through these it is now possible to separate living things into their molecular components with incredible resolution, determine their three-dimensional structures, follow the chemical reactions taking place in living things and determine how biomolecules perform the functions they do.

The power of a molecular perspective in the life sciences was aptly summarized by Francis Crick: 'Almost all aspects of life are engineered at the molecular level, and without understanding molecules we can only have a very sketchy understanding of life itself.' The aim of this book it to develop an understanding of life by describing molecules and how they work together in living systems. It does so using brief digestible introductions to 50 important ideas in biochemistry.

As with other titles in the series, this book does not attempt to give a systematic in-depth survey of biochemistry. Some of the entries in which the important ideas are presented explain key concepts or foundational principles that govern the behaviour of biological molecules. Others describe how molecules work individually or as part of larger systems to perform biological functions. Interspersed throughout are vignettes that present biochemistry as a human enterprise. Together these are intended to give you a sense for how life works at the molecular level and to help you see that there are always opportunities for bright and enterprising scientists to further our understanding of biochemistry.

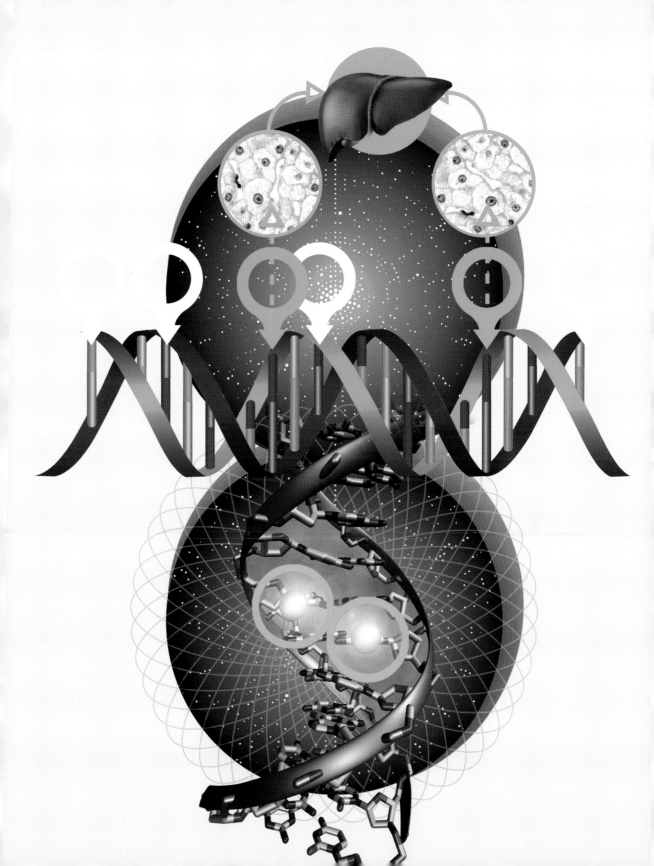

How this book works

Each entry is broken into several parts. The **30-second structure** and **3-second nucleus** summarize the main idea. In particular, the 30-second structure gives the body of that explanation while the 3-second nucleus summarizes it in a sentence or two. In contrast, the 3-second biographies and **3-minute system** explain how the main idea fits in a wider context. The **3-second biographies** introduce scientists who contributed to the idea's discovery or development. The 3-minute system expands on the main idea by explaining how it works in another context or how it applies in interesting situations.

Although each entry is capable of being read individually, it helps to begin with foundational principles before applying them to increasingly complex systems. The book begins by considering **biochemistry's context**, specifically its scope and the key physical principles that govern how molecules act in living systems. Next, the **building blocks** that make up life are described, followed by a more detailed look at the protein **machinery** that does much of the work of the cell. Then follow three sections that explore how collections of biomolecules work together to **harvest energy, make the molecules of life** and help complex multicellular organisms such as us **stay alive**. Finally, the way in which these processes can be disrupted for good or harm is explored in a section dealing with **sickness and health**. Fittingly, the final entry explores how biochemistry functions both as a way of making sense of life and transforming it, specifically through inexpensive and easy-to-use technologies for editing the information-bearing DNA molecules that shape the character of living things.

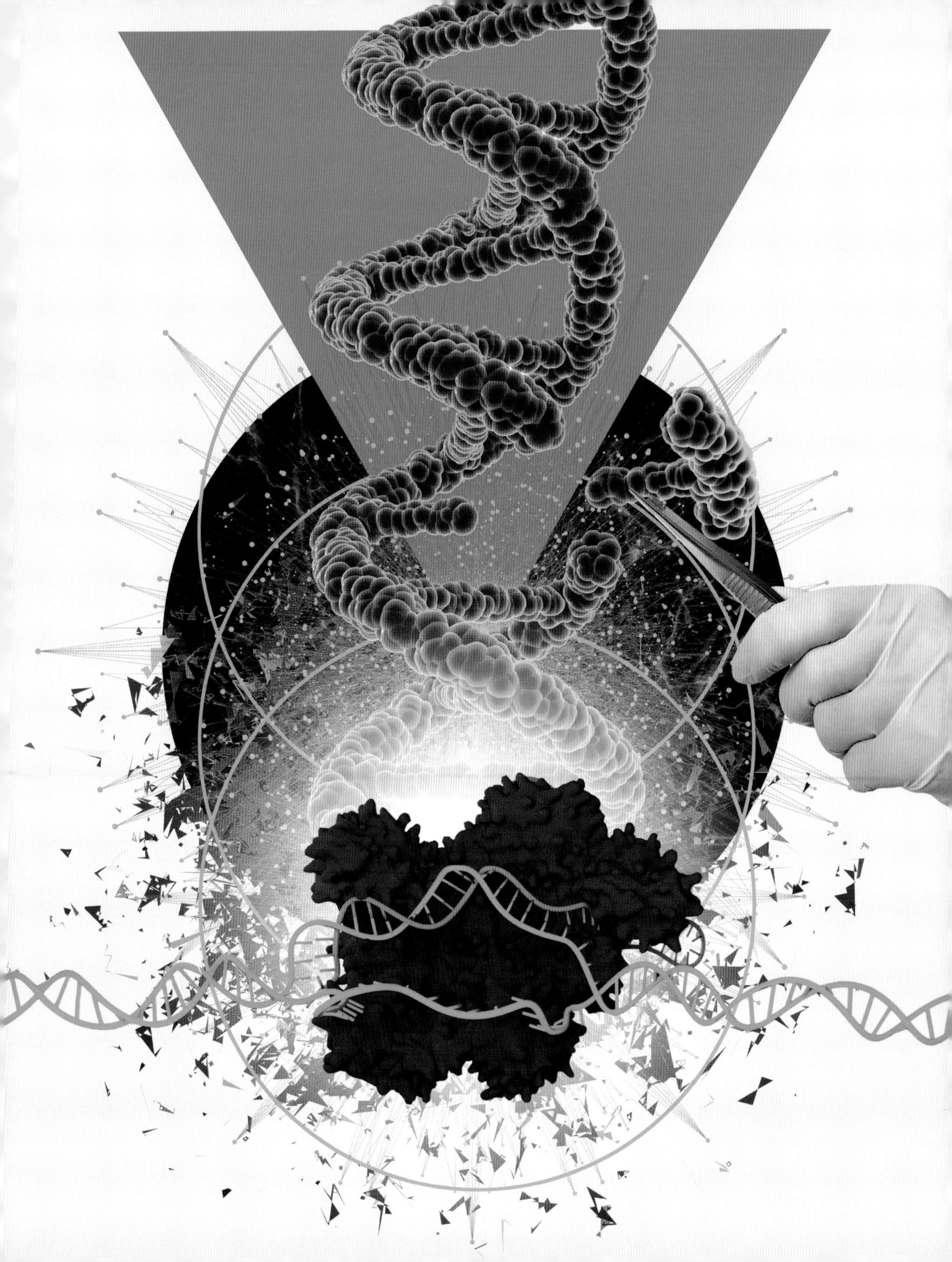

BIOCHEMISTRY'S CONTEXT

BIOCHEMISTRY'S CONTEXT
GLOSSARY

adenosine triphosphate (ATP) A nucleotide that serves as an energy currency, being formed in processes that release energy and broken down in processes that use energy.

adipocytes Specialized cells that store triglycerides or 'fats' in higher organisms.

biomolecule A molecule distinctively found in living systems.

bonds (chemical) Forces linking two atoms together in a molecule or solid.

buffer A mixture of an acid and base that maintains a constant acidity.

electrolytes Positively and negatively charged atoms or groupings of atoms, called ions, dissolved in water.

electron A negatively charged part of an atom. Attractions between electrons and positive nuclei are responsible for the forces between atoms.

endoplasmic reticulum An organelle consisting of a network of membranes where proteins and some lipids are made.

entropy A measure of the thermal energy in a system that is not able to do work. Since this energy is associated with uncoordinated motions, entropy is associated with the freedom a molecular system has to rearrange, sometimes loosely described as disorder.

eukaryotic cells Cells that contain smaller sacs, called organelles, which perform specialized functions.

glycolysis The breakdown of the carbohydrate glucose.

Golgi apparatus An organelle sometimes referred to as the cell's post office; it packages proteins for transport to other parts of the cell.

hydrophobic Something that appears to repel water since it doesn't stick tightly to it.

ion An atom or grouping of atoms possessing a positive or negative charge.

membrane A thin boundary (two lipids thick with a greasy inside and water-loving faces) that surrounds biological compartments such as cells and organelles.

mitochondria The power-plant organelles that produce most cellular ATP.

nucleotide A type of biomolecule consisting of one or more phosphates, a ribose or deoxyribose sugar and a flat nitrogen-containing group of atoms called a base. Nucleotides serve as information-bearing molecules, metabolic energy and electron carriers, and chemical messengers.

nucleus The organelle that contains genetic material and controls its transmission and expression.

organelle A small membrane-bound part of a eukaryotic cell that performs a specialized function.

organism A complete individual living thing.

prokaryotic cells Cells that consist mainly of a membrane-bound sac, with all the cellular apparatus inside.

LIFE'S SCALE

the 30-second structure

Biochemistry probes the workings
of life at chemistry's scale, concerning itself with
very small things such as atoms and molecules.
But since life is not so small, biochemists
investigate larger things, too. Biochemistry also
examines how atoms and molecules work
together as part of larger groupings, ranging
from a few biomolecules to thousands or even
many millions that work together in living
systems. For instance, even the smallest cell is
surrounded by a thin film consisting of around
10 million small greasy molecules called lipids.
Embedded within the film are larger proteins,
some of which have chain-like carbohydrate
molecules attached. Together, the lipid film,
embedded proteins and carbohydrates keep the
cell separate from its environment while enabling
it to take up nutrients, export waste and respond
to environmental signals. In plants and animals a
variety of molecular processes take place in and
between the specialized cells that comprise
structures such as tissues, organs and organ
systems. These processes enable organisms to
take up food, export waste, stay healthy,
reproduce and navigate surroundings ranging
from nearby cells to ecosystems. Finally, all
biochemical processes reflect life's dependence
on the energy derived from sunlight and nuclear
decay processes taking place within Earth's core.

RELATED TOPICS
See also
LIFE'S CURRENCY
page 74

BLOOD, BREATH & BALANCE
page 118

PROGENY & THE PILL
page 140

3-SECOND BIOGRAPHIES
CLAUDE BERNARD
1813–78
French physiologist who first
recognized that chemical
processes function at various
levels in living organisms

TADEUSZ REICHSTEIN
1897–1996
Polish-Swiss chemist who
isolated adrenal hormones
and identified links between
plant and insect biochemistry

30-SECOND TEXT
Stephen Contakes

3-SECOND NUCLEUS
Biochemistry involves the
study of structures – from
the tiny atoms and the
bonds that link them to
much larger organisms
and ecosystems.

3-MINUTE SYSTEM
Not all ecosystems are
external to an organism.
You, for example, are
not an organism but an
ecosystem. About 1–3 per
cent of the mass and
90 per cent of the cells
that make up your body
are bacteria. These serve as
a defence against harmful
pathogens and greatly
increase the metabolic
capabilities of your
intestinal tract. Without
them you would be unable
to digest the food you
eat or absorb the nutrients
you need to survive.

*From minuscule atoms
to large ecosystems,
biochemistry covers
a huge range of living
organisms on Earth.*

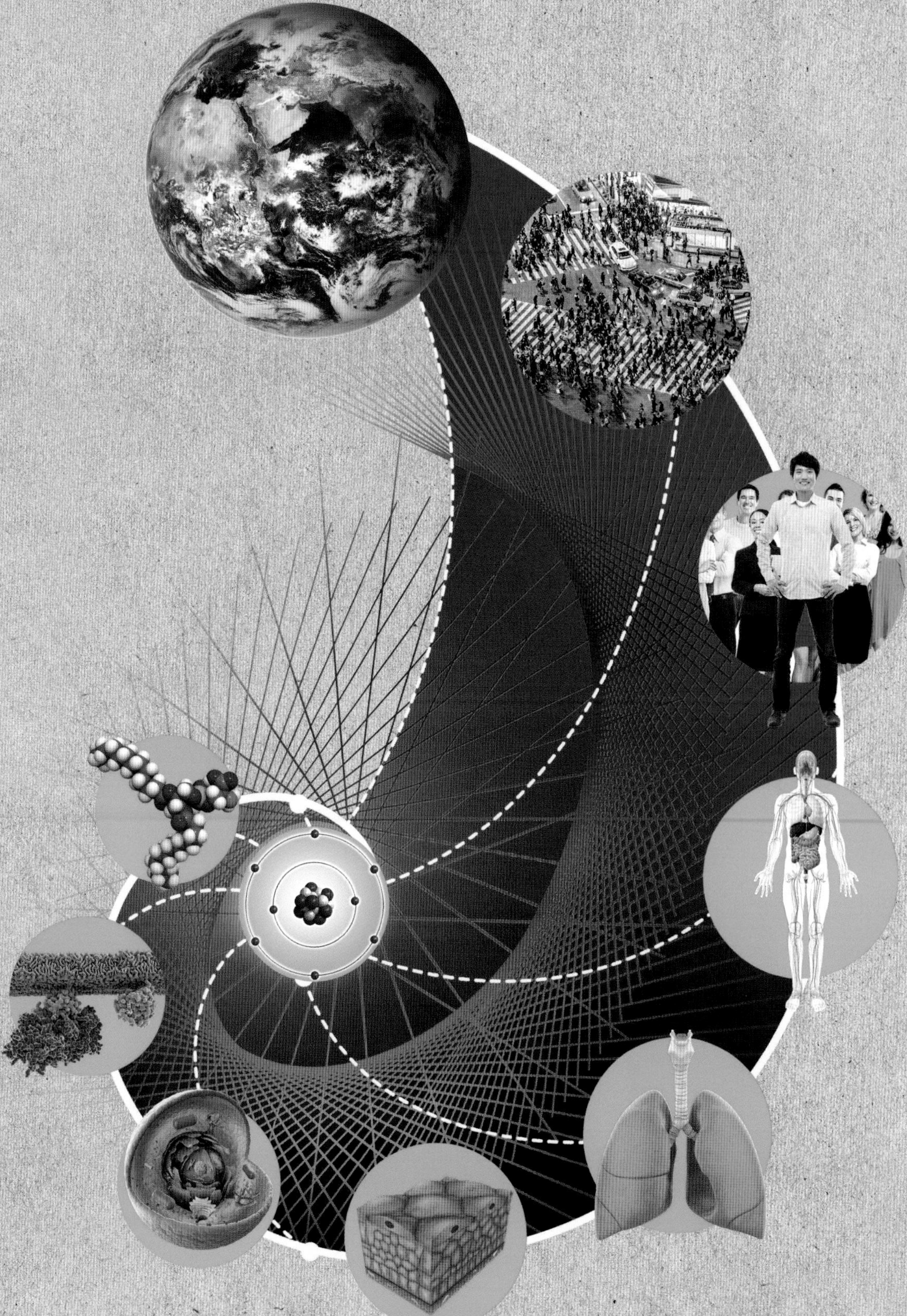

LIFE'S UNITS

the 30-second structure

The cell is the structural and functional unit of all living organisms. Imagine every cell like a house with its outer wall made of a lipid membrane. There are cells, called eukaryotic cells, with various membrane structures inside – the organelles (like rooms in a house) – performing cellular functions and bathing in a gelatinous fluid, the cytoplasm. The most famous organelle in eukaryotic cells is the nucleus, where the genetic material is hosted. Other organelles include the mitochondria (the energy factories of the cell and the place where the inhaled oxygen ends up in the body), the endoplasmic reticulum (a protein factory and transport network), the Golgi apparatus (where molecules synthesized in the cell are packaged and shipped to other parts of the cell) and the lysosomes and peroxisomes (the cell's cleaners). Cells lacking organelles (like houses without rooms) are called prokaryotic and were the first to appear on Earth, with evolution leading to more complex cells, the eukaryotic ones. Animals (including us humans) and plants are multicellular organisms made of eukaryotic cells, whereas unicellular organisms can be either prokaryotic, such as the bacteria, or eukaryotic, such as yeast. The size and weight of a living organism depend on the number and size of its cells.

3-SECOND NUCLEUS
The cell is the basic unit of life; in fact, there are living organisms that are nothing more than a single cell.

3-MINUTE SYSTEM
In higher organisms, life begins with the generation of a first cell by the union of the parents' two genetic cells (egg and sperm in animals). The whole organism is then formed thanks to the unique gift of cells to divide and multiply. A human life starting with a single cell develops into an organism with about 40 trillion cells. These differ in shape and function, making tissues and organs thanks to cells' ability to differentiate into specific types (muscle, liver, fat and so on).

RELATED TOPICS
See also
LIPIDS
page 40

LIFE'S LETTERS: NUCLEOTIDES
page 46

MEMBRANES, MESSENGERS & RECEPTORS
page 66

MAINTAINING THE CODE
page 102

3-SECOND BIOGRAPHIES
MATTHIAS SCHLEIDEN & THEODOR SCHWANN
1804–81 & 1810–82
German botanist and physiologist who introduced the idea that cells are the basic unit of life

RUDOLF VIRCHOW
1821–1902
German pathologist who expanded cell theory to explain that all cells come from pre-existing cells

30-SECOND TEXT
Anatoli Petridou

The building blocks of life: a eukaryotic cell (top) and a prokaryotic cell (bottom).

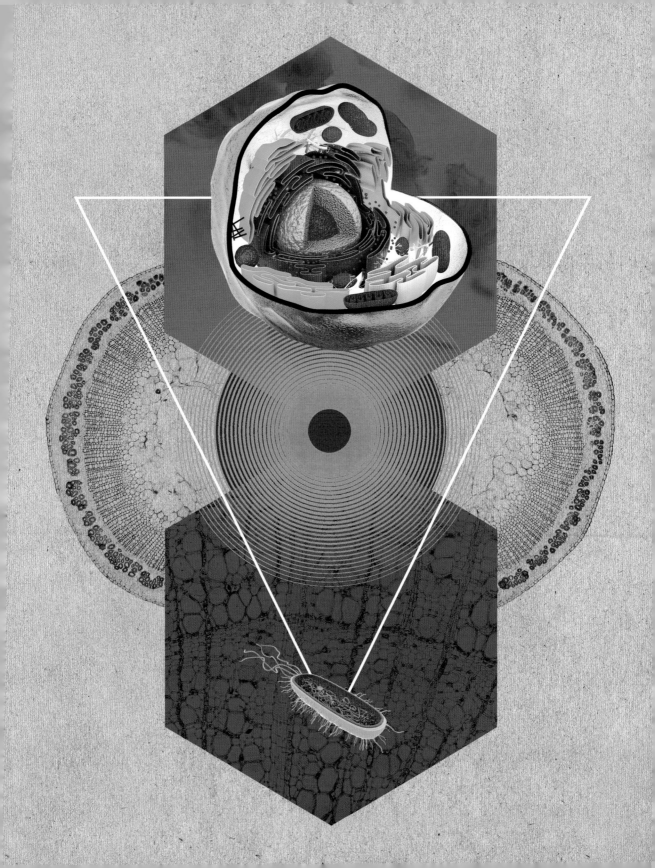

LIFE'S POWER

the 30-second structure

The smallest cell in the human body ultimately gets its power to live from the largest object in the solar system: the Sun. As the energy flows from the Sun to humans, it is neither created nor destroyed, but is transformed, passing from the realm of astrophysics to the realm of biochemistry. First, the Sun's mass is transformed into radiant light through nuclear fusion, as atomic nuclei collide and release energy according to the equation $E=mc^2$. The light travels to Earth, is captured by colourful pigments in a plant and is used to push electrons in a process called photosynthesis. Photosynthesis synthesizes two types of chemical bonds: carbon–carbon bonds, such as those in a sugar molecule, and oxygen–oxygen bonds in the air. An animal that eats plants and inhales oxygen digests its sugar through glycolysis (literally, 'sweet breaking'), trading carbon–carbon bonds for more stable carbon–oxygen bonds. This powers the formation of phosphate bonds in the molecule adenosine triphosphate (ATP). The phosphate bonds in ATP can be traded for all sorts of other bonds, building many types of molecules that let the body move, breathe and find more sugar. Each movement means that more energy flows from the Sun through the body in a cascade of transformation.

3-SECOND NUCLEUS
Sunlight is transformed into chemical bonds that power life through the biochemical reactions connecting all living things.

3-MINUTE SYSTEM
Not all life gets its energy from the Sun. Simple organisms can extract small amounts of chemical energy from inorganic compounds in the ocean or rocks containing sulphate, iron or even manganese. They do this by moving electrons from an unstable location to a more stable location, harvesting some of the energy released and trading it for the chemical bonds they need to live.

RELATED TOPICS
See also
LIFE'S CURRENCY
page 74

USING BIG CARBS
page 76

USING SMALL CARBS
page 78

BURNING IN REVERSE:
PHOTOSYNTHESIS
page 94

3-SECOND BIOGRAPHIES
FRITZ LIPMANN
1899–1986
German-American biochemist who established ATP as the cell's energy currency

MELVIN CALVIN
1911–97
American biochemist who discovered the Calvin cycle of reactions that turn carbon dioxide into glucose

30-SECOND TEXT
Ben McFarland

Sunlight is the ultimate source of energy for the chemical reactions that power life.

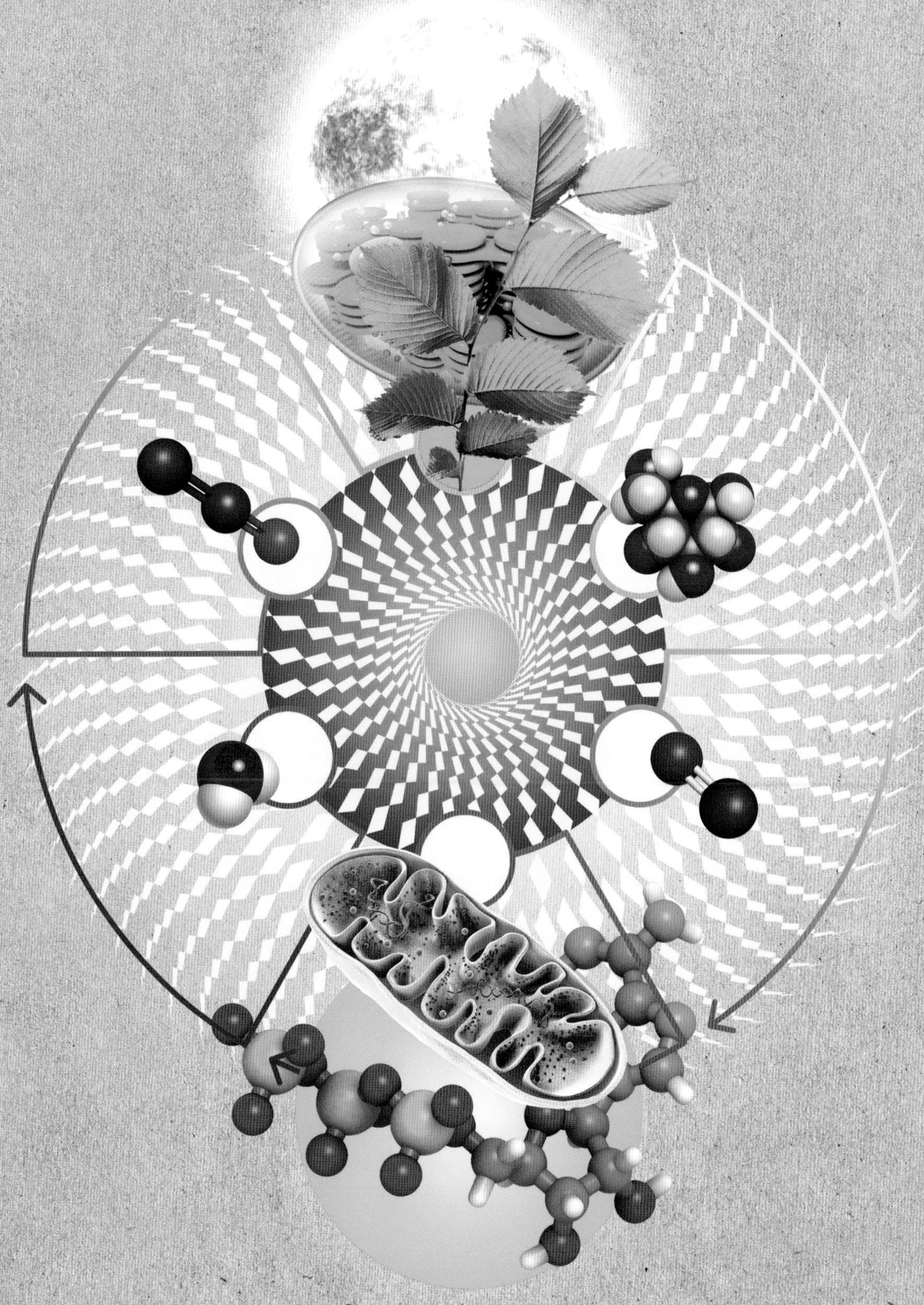

LIFE'S RULES

the 30-second structure

3-SECOND NUCLEUS
Just try to disobey: life
is a constant struggle
against the second law
of thermodynamics, and
entropy always wins.

3-MINUTE SYSTEM
The assembly of
virus particles seems
counterintuitive to the
second law. Viruses
form when several small
proteins called subunits –
sometimes hundreds and
even thousands – assemble
into a large particle. This
restricts their motions
and reduces their entropy.
This entropy-lowering
process is counterbalanced
by dehydration: once
the virus forms, water
molecules that were stuck
to the surface of the
subunits are liberated,
giving the overall process
an increase in entropy, thus
satisfying the second law
of thermodynamics.

In the branch of chemistry
dealing with energy, the second law of
thermodynamics states that the entropy of the
universe is always increasing. Simply, entropy is
a measure of randomness throughout a system.
More precisely, entropy is a measure of the
thermal energy not able to do work, arising as
energy becomes more evenly distributed as a
system gains more freedom to reconfigure itself
– described as a state of 'disorder'. At its core,
concepts of entropy are based on human
experience: consider ice melting. Ice is a well-
ordered crystalline system of water molecules.
When ice is heated, the water molecules
eventually gain enough energy to break the
restricted crystalline structure, forming fluid
water. In this example, the increase in heat
energy enables the system to transition from
being localized and highly ordered to spreading
out. There is no useful work resulting from
dispersal; indeed, 'energy gone to waste' is a
term used by Rudolf Clausius to describe the
concept of entropy. This has significant
implications for life, as organisms are highly
ordered molecular constructs. Life is a constant
struggle against the tendency of entropy to
disperse molecular systems, one in which survival
involves coupling the production of complex
biomolecules to entropy-increasing processes.

RELATED TOPICS
See also
LIFE'S ORGANIZING FORCE
page 28

3-SECOND BIOGRAPHIES
SADI CARNOT
1796–1832
French physicist who provided
theories of heat engines used
by Clausius to develop entropy

RUDOLF CLAUSIUS
1822–88
German physicist who coined
the term 'entropy'

JOSIAH WILLARD GIBBS
1839–1903
American scientist who
established how entropy
determines the course of
chemical reactions

30-SECOND TEXT
Andrew K. Udit

*Melting ice releases
water molecules from
a structured solid;
this 'molecular
relaxation' results in
increased entropy.*

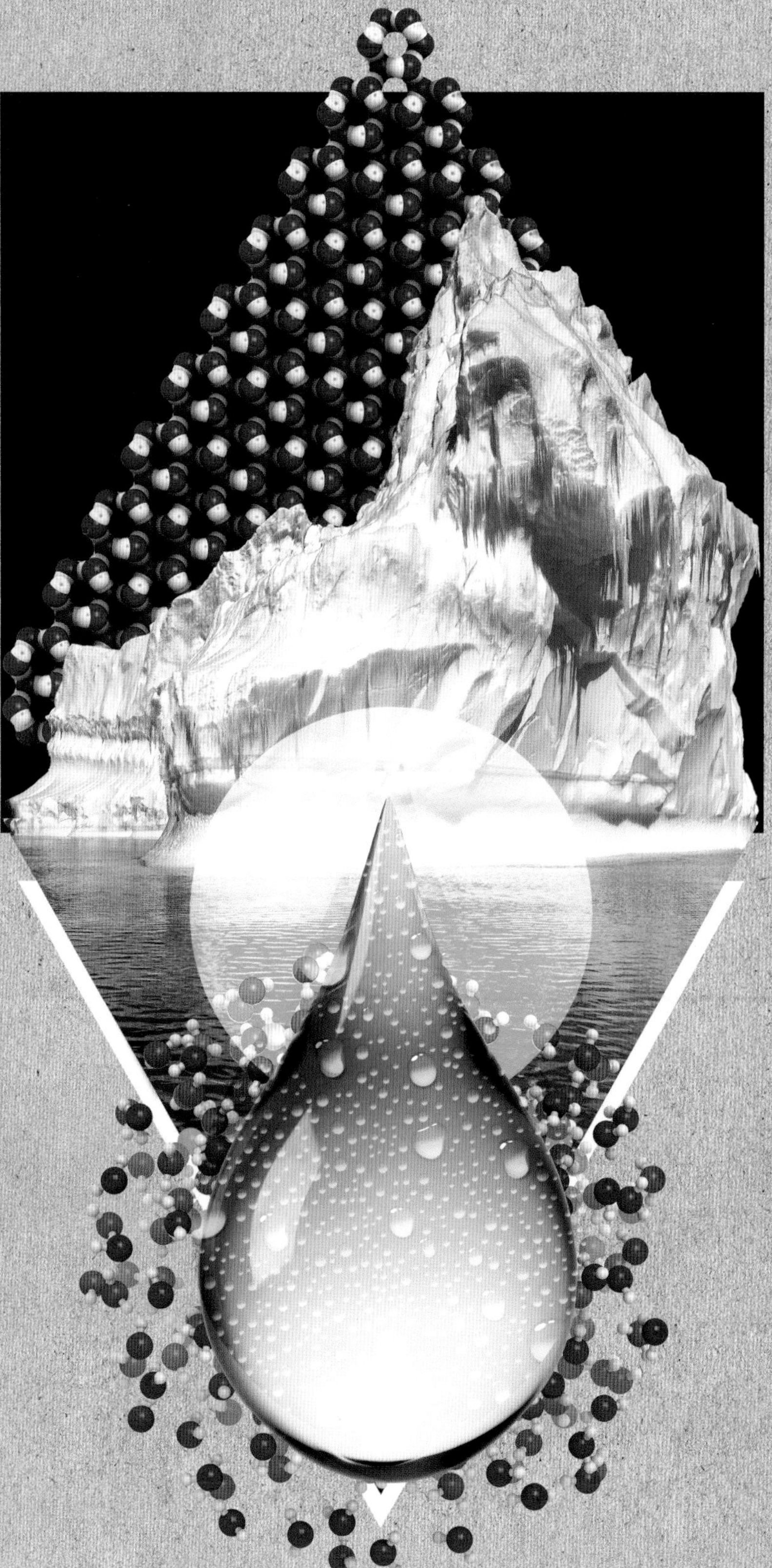

LIFE'S MATRIX

the 30-second structure

Life itself is poised between
order and chaos, and it is composed of a phase
of matter poised between the order of a solid
and the chaos of a gas: liquid water. Water's
formula, H_2O, uniquely combines two opposing
properties. It is simple and therefore abundant,
but its hydrogen and oxygen atoms share
electrons in chemical bonds unequally, giving a
structure with electrically positive and negative
ends called an electric dipole. The positive and
negative ends of a water molecule's dipole
attract the oppositely charged ends of other
water molecules while water's hydrogen can
hold together two oxygen atoms by forming
O–H---O bridges called hydrogen bonds. Thus,
water molecules stick together tightly like
tiny magnets. At Earth temperatures, these
attractions keep the water molecules in a cell
adjacent but moving constantly, changing
partners like dancers at a Victorian ball. The
negative end of water's dipole can stick to
positively charged ions, enabling water to
dissolve things such as salt. These dissolved ions
make the ocean salty, and life evolved to fit this
environment. As life emerged from the ocean
and moved around on dry land, it retained its
chemical need for flowing water and salty ions,
and the blood filled this need. Therefore, the
blood has a similar composition to the ocean,
and each animal contains its own small sea.

3-SECOND NUCLEUS
Water's unique properties
and abundance on this
planet make it an excellent
basis for life.

3-MINUTE SYSTEM
An overheated body cools
itself by producing salty
sweat. This lost salt can be
replenished with sports
drinks containing
electrolytes. The first
sports drinks were
developed by doctors at
the University of Florida
to help football players
endure subtropical
heat at practice sessions.
Being based on a scientific
hypothesis rather than
the experience of taste, the
first formulations of these
drinks tasted terrible, but
later versions incorporating
lemon juice and sweetener
were more palatable.

RELATED TOPICS
See also
LIFE'S ORGANIZING FORCE
page 28

BLOOD, BREATH & BALANCE
page 118

ELECTROLYTE BALANCE
page 128

3-SECOND BIOGRAPHIES
T. S. MOORE & T. F. WINMILL
fl. 1912
British chemists who first
identified the hydrogen bond

JAMES ROBERT CADE
1927–2007
American physician who
formulated the electrolyte-rich
sports drink

30-SECOND TEXT
Ben McFarland

*Water molecules
adhere to each other
with hydrogen bonds.*

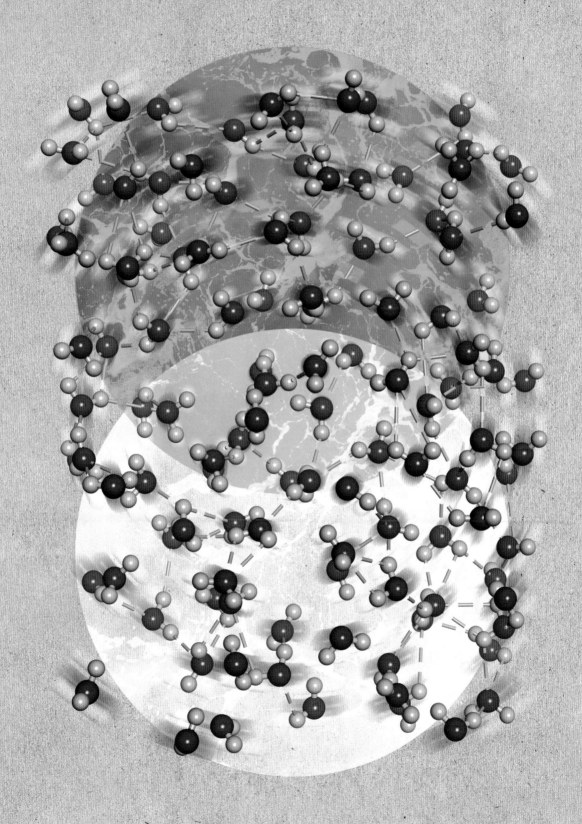

LIFE'S GIVE & TAKE

the 30-second structure

3-SECOND NUCLEUS
Not too acidic, not too basic, but just right – buffers maintain this balance so that life can function in the necessary sweet spot.

3-MINUTE SYSTEM
The acidity or basicity of solutions is rated by the concentration of H_3O^+ using the pH scale. A pH of 7 is typically regarded as neutral – that is, equal amounts of H_3O^+ and OH^-; pH values higher than that are basic, lower ones acidic. Many biological systems are optimized to function in a narrow pH range. For instance, the degradation of proteins in the stomach requires an acidic pH.

A sample of pure water is more than simply H_2O. Water undergoes an auto-ionization process that results in the transfer of an H^+ ion from one water molecule to another, yielding OH^- and H_3O^+ ions: $2H_2O \rightarrow H_3O^+ + OH^-$. An acid is anything that will increase the amount of H_3O^+ in solution, while a base will increase the amount of OH^-. It is this balance between H_3O^+ and OH^- that qualifies a solution as either being acidic (more H_3O^+ than OH^-) or basic (more OH^- than H_3O^+). Maintaining the right ratio of H_3O^+ to OH^- is critical for many biological processes, and it does not always have to be balanced evenly: while blood requires an almost even ratio (slightly basic), the stomach needs a highly acidic environment to digest food. Buffers are used to maintain the appropriate ratio of H_3O^+ to OH^-. A buffer is a mixture of related acid and base components in similar amounts. It resists changes to the H_3O^+ to OH^- ratio by absorbing or releasing these two ions as needed. If an acid is added to a buffer, the base component can 'absorb' it, thus maintaining the H_3O^+ to OH^- ratio; an added base is similarly absorbed by the acid component.

RELATED TOPICS
See also
LIFE'S MATRIX
page 22

LAWRENCE HENDERSON
page 26

BLOOD, BREATH & BALANCE
page 118

3-SECOND BIOGRAPHIES
SVANTE AUGUST ARRHENIUS
1859–1927
Swedish chemist who first defined acids and bases in aqueous solution using H^+ and OH^-

SØREN PETER LAURITZ SØRENSEN
1868–1939
Danish chemist who defined the concept of pH

30-SECOND TEXT
Andrew K. Udit

Buffers perform a biological balancing act in order to maintain a specific acid–base ratio in the body.

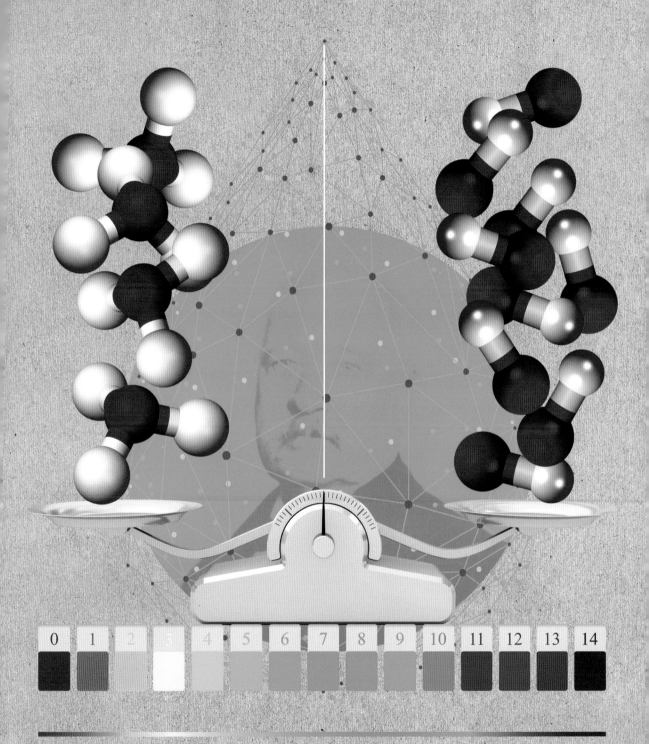

3 June 1878
Born in Lynn,
Massachusetts, USA

1898
Graduates from Harvard
College

1902
Receives a medical degree
from Harvard Medical
School and begins studies
in biological chemistry
in Franz Hofmeister's
laboratory at the
University of Strasbourg

1904
Joins the faculty of
Harvard Medical School,
where he will spend the
remainder of his career

1908
As part of his efforts to
understand how blood's
acidity is regulated,
publishes an equation
that describes the pH
of a buffer solution

1913
Publishes *The Fitness
of the Environment*,
interpreting biological
evolution in the light
of geological and
environmental chemistry

1917
Publishes *The Order of
Nature*, integrating
astrophysics and cosmic
evolution with the
chemistry of three
elements (carbon,
oxygen and hydrogen)

1924
Becomes first president
of the History of Science
Society

1928
Publishes *Blood: A Study
in General Physiology*,
using innovative
mathematical diagrams
called nomograms to
explain how blood
works in different
parts of the body

1935
Publishes *Pareto's
General Sociology:
A Physiologist's
Interpretation*, which
interprets Pareto's
sociological observations
in terms of physical
chemistry parameters
such as temperature
and pressure

10 February 1942
Dies in Cambridge,
Massachusetts, USA

LAWRENCE HENDERSON

Lawrence Joseph Henderson, one of the leading biochemists of the early twentieth century, applied the mathematical rigour of physical chemistry to life's interrelated systems, including blood chemistry and ecological fitness. Born in 1878 in Massachusetts, Henderson soon showed an aptitude in mathematics and physics that led to his enrolment in Harvard College when he was 16 years old. He earned a medical degree from Harvard Medical School in 1902 and spent two years studying haemoglobin chemistry in Germany, after which he returned to Harvard to teach and study biological chemistry, working as a professor from 1919 until his death in 1942. During this time he worked in many fields: chemistry, biology, medicine, philosophy, history and sociology.

Henderson wondered how blood pH stayed the same, being 'buffered' to remain at neutral pH, despite constant fluctuations of acid and base in the environment. He found an equation that connected the buffering strength of a molecule with a chemical constant (the acid dissociation constant). This became the Henderson–Hasselbalch equation, still widely used by chemists and biologists today. Henderson found that the chemical constants of phosphate and carbonate together provide a powerful buffer that keeps blood pH the same.

Henderson drew philosophical implications from these scientific findings. Blood's special buffering chemistry relies on the fixed chemical properties of elements such as carbon and oxygen. Organisms need to eat and excrete these elements while maintaining constant chemical levels inside, leading to an essential chemical relationship between the organism and its environment. The properties and availability of molecules made from these elements (especially water and carbon dioxide) therefore fit the needs of living beings, as Henderson wrote in *The Fitness of the Environment* (1913).

While Charles Darwin focused on how organisms fit their environment, Henderson focused on what kind of environment allows diverse organisms to prove their fitness. Henderson noticed that only a few elements can make environments suitable for a high structural and chemical diversity of organisms. Because water and carbon dioxide have suitable chemical properties for life and are widely available, they are likely to be universally useful to make biochemical creatures that are free to self-regulate, develop and evolve.

Alongside his teaching and studies, he combined the fields of sociology and medicine in his 1935 book *Pareto's General Sociology: A Physiologist's Interpretation* and his 1941 essay 'The Study of Man'. His work thrived at the intersection of diverse fields and established patterns of thinking about systems that prepared the way for interdisciplinary investigations today.

Ben McFarland

LIFE'S ORGANIZING FORCE

the 30-second structure

Oil and water don't mix – but why not? When you see drops of oil gather in water, it's tempting to guess that the oil is forming strong bonds with itself. That guess doesn't account for the most important part of the system: the water. Water molecules next to oily molecules cannot form stabilizing hydrogen bonds and are frozen in what scientists call a 'frustrated' position. When two oil drops join, the frustration is relieved as these water molecules are unfrozen and escape to form hydrogen bonds with many other water molecules. Because the oil appears to avoid the water, this is called the hydrophobic effect. Hydrophobic interactions hold many oily or half-oily molecules together in water: cell and viral membranes, protein interiors, DNA bases and the fat globules of seeds and human adipocytes (fat cells) are all held together by this automatic effect. Hydrophobic interactions can be targeted with half-hydrophobic molecules such as soaps. Soap dissolves oily structures by bridging the oil and water molecules, disrupting the hydrophobic interactions. Therefore, something as simple as washing your hands can break open the membranes around encapsulated viruses.

RELATED TOPICS
See also
LIFE'S RULES
page 20

LIFE'S MATRIX
page 22

LIPIDS
page 40

SHAPE: PROTEIN STRUCTURE
page 56

3-SECOND BIOGRAPHIES
IRVING LANGMUIR
1881–1957
American chemist who won the 1932 Nobel Prize for work on the chemistry of oil films

HENRY S. FRANK
1902–90
American chemist who proposed how entropy could drive the aggregation of oil molecules by the increase of water's entropy

30-SECOND TEXT
Ben McFarland

The hydrophobic effect causes the oily ends of molecules like those in soap to clump together into spherical droplets in water.

3-SECOND NUCLEUS
The hydrophobic effect explains how oily molecules collect together in water; it holds cell membranes and proteins together, organizing life.

3-MINUTE SYSTEM
Flat carbon surfaces are hydrophobic, but when combined with a microscopically rugged surface the result is a superhydrophobic surface that repels water even more. Liquid water will form a near-perfect sphere on a superhydrophobic surface. Some cacti use tiny superhydrophobic cones to collect water from fog. Some desert beetles do the same with superhydrophobic surfaces on their backs. Other insects shatter raindrops with their hydrophobic carapaces, deflecting the force of the falling water so that they are not flattened by every passing storm.

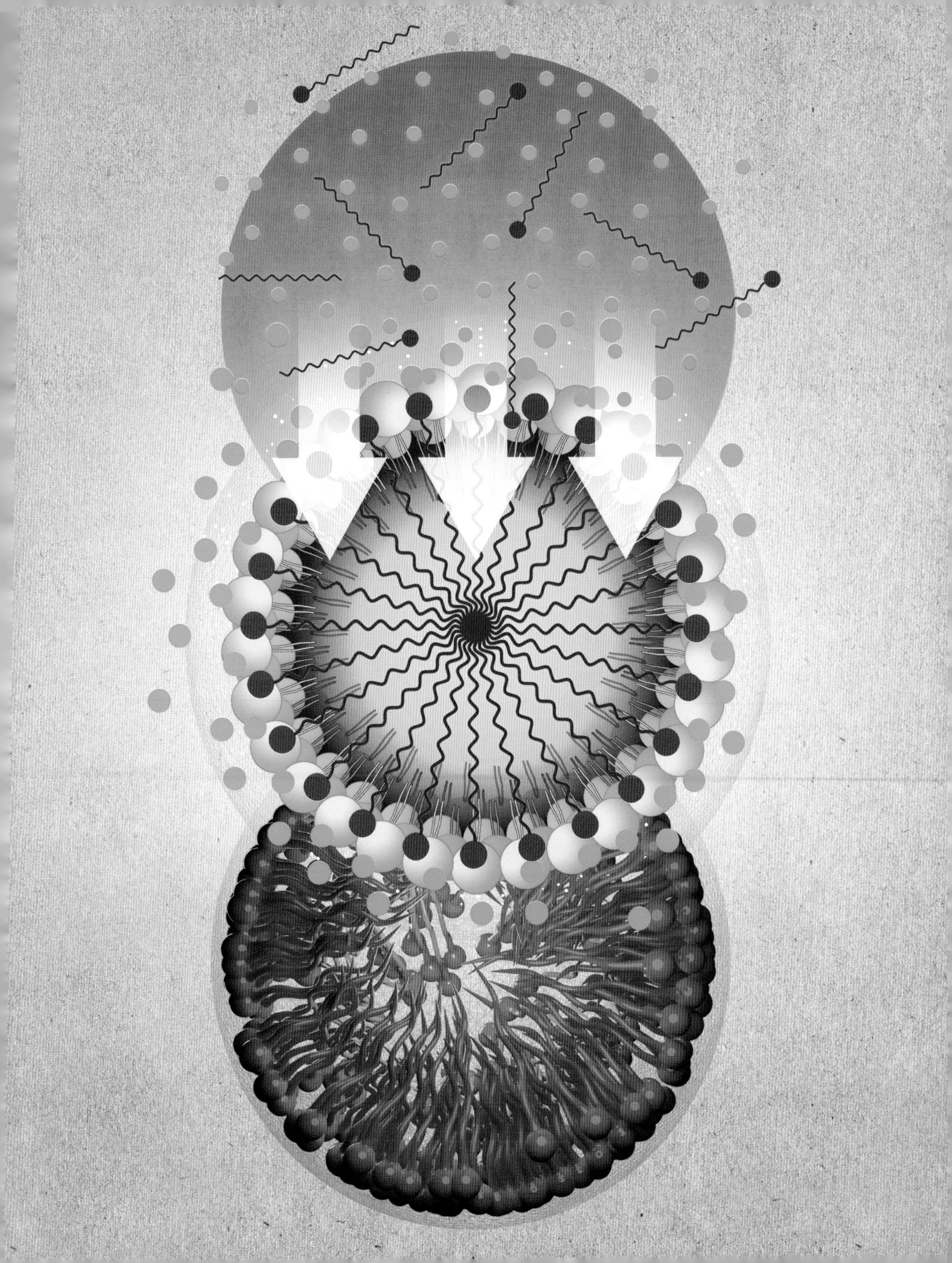

BUILDING BLOCKS

chiral Molecules which, like your right and left hands, are not superimposable with their mirror images.

covalent bond Chemical bond formed by sharing electrons between atoms, as opposed to an ionic bond, which involves attractions between positively and negatively charged ions.

disaccharides A type of carbohydrate that consists of two monosaccharide rings linked together. Examples include sucrose (table sugar) and lactose (milk sugar).

electron A negatively charged part of an atom. Attractions between electrons and positive nuclei are responsible for the forces between atoms.

glycogen The polysaccharide which stores glucose units in muscle and liver in a bush-like structure that facilitates their rapid breakdown.

hydrophilic Something that can form strong interactions with water.

hydrophobic Something that appears to repel water since it doesn't stick tightly to it.

hydroxyl Part of a molecule in which a hydrogen is bound to an oxygen so that the two comprise an O-H or OH. An example of a hydroxyl group occurs in ethanol, CH_3-O-H, which is produced by the breakdown and fermentation of sugars.

ion An atom or grouping of atoms possessing a positive or negative charge.

monomer A unit that is hooked together with other identical or similar ones to give long chain molecules, called polymers.

monosaccharides Carbohydrates containing a single chain of carbons in a linear or ring form that cannot be split into smaller units by reaction with water.

nucleic acids DNA and RNA polymers of nucleotides. These are called nucleic acids because they are derivatives of phosphoric acid mainly residing in and around the cell nucleus; they are involved in the storage and expression of genetic information.

nucleobase The variable flat part of a nucleotide. The order in which they occur along a nucleic acid chain encodes genetic information.

polymer A molecule made by joining smaller units, called monomers, into a long chain.

polysaccharides Carbohydrate polymers made by linking many monosaccharide units together.

ribosome An associated collection of RNA and protein that helps link amino acids into proteins.

synthesis A chemical reaction in which something of interest is made. In biochemistry it usually refers to the formation of larger or more complex molecules from smaller ones.

supramolecular complex A collection of molecules held together without covalent bonds that together comprise a larger unit.

triglycerides Lipid major component of animal fats and plant oils that consists of three fatty acids linked to a single glycerol unit.

ELEMENTS

the 30-second structure

Living organisms contain a relatively small number of chemical elements. Fewer than 25 of the more than 90 naturally occurring elements are essential to organisms. The four most abundant elements in living matter – carbon, hydrogen, oxygen and nitrogen – account for more than 99 per cent of the mass of most cells. The abundance of hydrogen and oxygen is due, in part, to the abundance of water in living organisms. Your body contains approximately 70 per cent water by weight. Carbon, hydrogen, oxygen and nitrogen are important constituents of the four major types of biomolecules – lipids, carbohydrates, nucleic acids and proteins. These non-metal elements form strong covalent bonds where electrons are shared within the bonds. Other important elements in living organisms include the non-metals sulphur and phosphorus as well as the metals calcium, sodium, potassium, magnesium, zinc, iron and copper. Sulphur is a part of some amino acids in proteins, and phosphorus is in the nucleic acids that form DNA. Phosphorus is also a constituent of adenosine triphosphate (ATP), the cell's main energy currency. ATP must bind to magnesium in order to be biologically active. The protein haemoglobin in red blood cells contains an iron atom that binds oxygen in the lungs and transports it to your muscles and other tissues.

3-SECOND NUCLEUS
Life depends primarily on a few chemical elements, although many others have essential functions as well.

3-MINUTE SYSTEM
The sodium-potassium pump is a protein machine found in our cell membranes. It pumps sodium ions out of, and potassium ions into, the cell in a process that is powered by ATP. The pump generates a gradient of ions that is used to maintain cellular volume and to transmit signals between nerve cells in the brain. Approximately one-third of the ATP made by our cells is used to power the sodium-potassium pump.

RELATED TOPICS
See also
ATP FROM AIR
page 86

BURNING IN REVERSE:
PHOTOSYNTHESIS
page 94

BRINGING IN THE NITROGEN
page 96

BIOCHEMISTRY ON THE BRAIN
page 124

3-SECOND BIOGRAPHIES
ANTOINE LAVOISIER
1743–94
French chemist who recognized that breathing involves burning substances with elemental oxygen to generate energy

FRIEDRICH WÖHLER
1800–82
German chemist who showed that living systems obeyed the ordinary rules of chemistry by synthesizing urea, a component of urine, from an inorganic salt

30-SECOND TEXT
Kristi Lazar Cantrell

Many chemical elements are essential for life.

H											C	N	O
Na	Mg										P	S	Cl
K	Ca		V	Cr	Mn	Fe	Co	Ni	Cu	Zn		Se	Br
				Mo									I
				W									

BONDS: LIFE'S HANDEDNESS

the 30-second structure

No other element is so small and versatile as carbon. Carbon's ability to form four strong, directional chemical bonds to other atoms makes it an essential building block for the complex, stable structures and long chains required for life's diversity. Carbon can bond the four atoms in two different ways, making two different 'chiral' shapes: a left-handed (L) and right-handed (D) version. Almost all biomolecules exist in only one of the possible forms. For instance, the amino acid building blocks of proteins could theoretically adopt either structure, but for some reason all known living things use only left-handed (L) amino acids. The left-handed version of a molecule looks exactly the same to most chemical methods of separation and analysis, with two major exceptions: the different forms of chiral molecules form differently shaped crystals and twist polarized light in different directions. Chiral protein catalysts selectively work with amino acids of a particular handedness. Simulations predict that proteins built from amino acids of only one handedness have more consistent structures than proteins built from a mixture of both. Yet we don't know why left-handed amino acids were selected. Perhaps a primordial flip of a coin chose left-handed amino acids, and this choice was inherited by all life on Earth.

3-SECOND NUCLEUS
Carbon can form four bonds to give three-dimensional structures that can exist as left-handed or right-handed forms, but life has selected only the left-handed version, at least here on Earth.

3-MINUTE SYSTEM
Chiral molecules twist polarized light on the planetary scale. The green light leaving the Earth has a distinctive twist in it from all the proteins made by plants. If this signal was detected in the light from an alien planet, it would be evidence for life. This signal has been detected in the light from our planet, in the 'earthshine' reflecting off the Moon during an eclipse, showing this works in principle, although the signal is very faint.

RELATED TOPICS
See also
WORKFORCE: AMINO ACIDS & PEPTIDES
page 48

PROTEIN FOLDING & AGGREGATION
page 54

POOLING THE AMINO ACIDS
page 98

3-SECOND BIOGRAPHIES
JEAN-BAPTISTE BIOT, AUGUSTIN FRESNEL & AIMÉ COTTON
1774–1862, 1788–1827 & 1869–1951
French scientists who developed circular dichroism, a technique that detects chiral molecules

LOUIS PASTEUR
1822–95
French scientist who separated two chiral forms of tartaric acid from crystals called 'wine diamonds'

30-SECOND TEXT
Ben McFarland

All amino acids have a chiral handedness that twists polarized light.

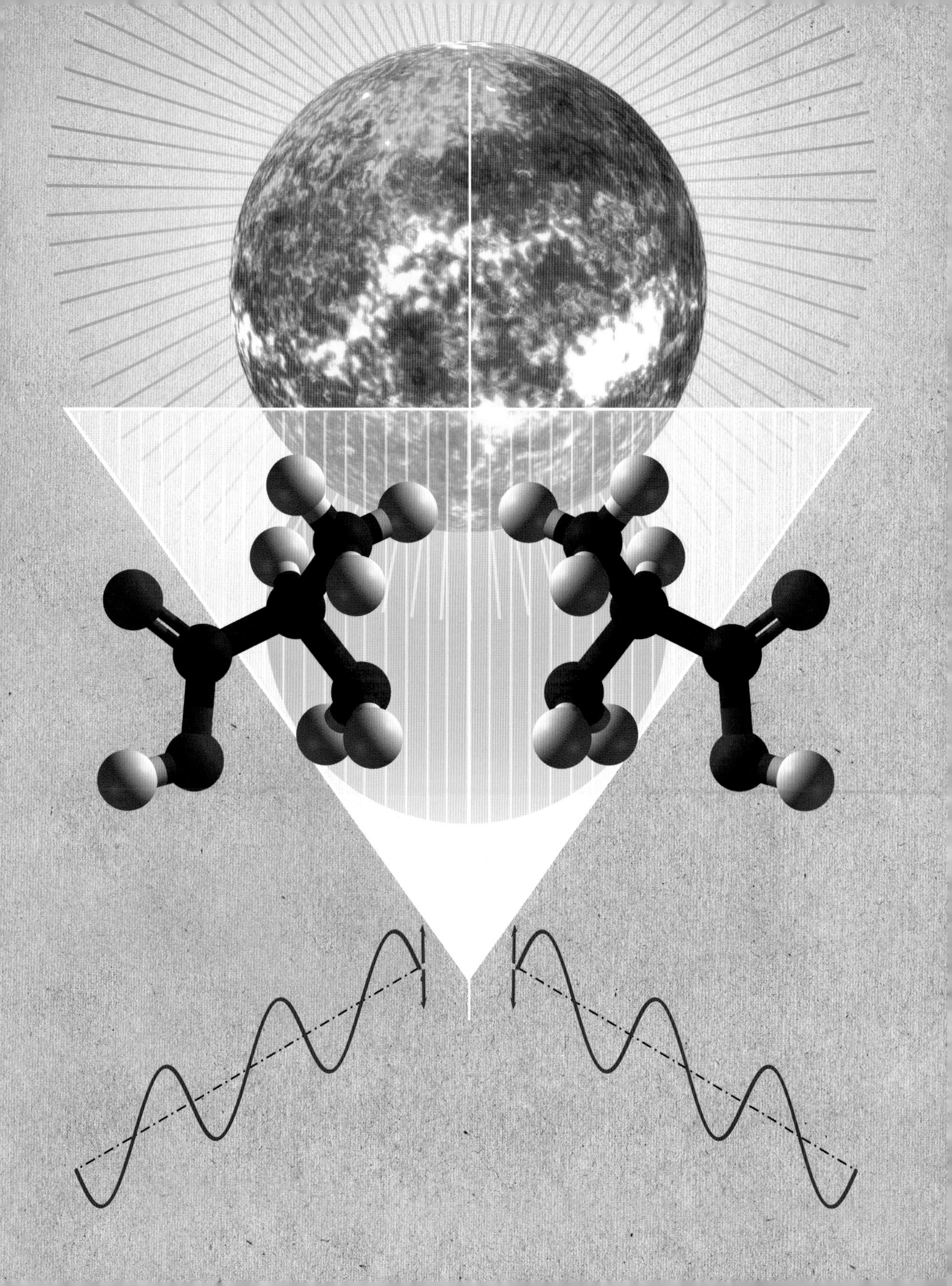

CHAINS: BIOLOGICAL POLYMERS

the 30-second structure

3-SECOND NUCLEUS
Many of the carbohydrates, amino acids and nucleic acids in our cells are formed by linking small molecule modules together into chains called polymers.

3-MINUTE SYSTEM
More than half of the carbon in the biosphere consists of glucose in the form of cellulose. It is mostly found in plant cell walls, in which its straight chains associate to form tough fibres. In contrast, glycogen formed in muscles and the liver after a meal contains coiled chains of glucose, some of which act as three-way linkers. These additional coils branch off the glucose chain, giving glycogen a bushy appearance.

Of the four major types of biomolecules, carbohydrates, amino acids and nucleic acids form long, chain-like structures called polymers. In biological polymers, individual modules – called monomers or residues – are linked together via a process called condensation in which the atoms lost as the monomers join together are released in the form of water. The monomers in biological polymers are arranged in specific ways. Among the many ways to join glucose molecules, the thousands that form the chains of cellulose are all linked together in one that results in straight chains rather than kinked ones. In other biological polymers the monomers are linked in particular sequences, just as the letters of this sentence are. This allows polymers of nucleic acids, such as DNA and RNA, to carry biological information and polymers of amino acids to perform a dazzling array of functions. Sometimes polymer chains associate with one another to form larger agglomerations called supramolecular complexes. The ribosome is a supramolecular complex of protein and RNA that links amino acids together to form new protein chains. At the end of their lifetime these protein chains are broken down in turn by another supramolecular complex, the proteasome, which is itself comprised of associated protein chains.

RELATED TOPICS
See also
SWEETNESS: CARBOHYDRATES
page 42

LIFE'S LETTERS: NUCLEOTIDES
page 46

WORKFORCE: AMINO ACIDS
& PEPTIDES
page 48

READING THE BOOK
page 106

3-SECOND BIOGRAPHIES
ANSELME PAYEN
1795–1871
French chemist who first isolated cellulose from plant matter and determined its chemical formula

CLAUDE BERNARD
1813–78
French physiologist who first isolated glycogen from liver tissue and determined its physical and chemical properties

30-SECOND TEXT
Kristi Lazar Cantrell

Chains of glucose form cellulose polymers – the tough fibres found in plant cell walls.

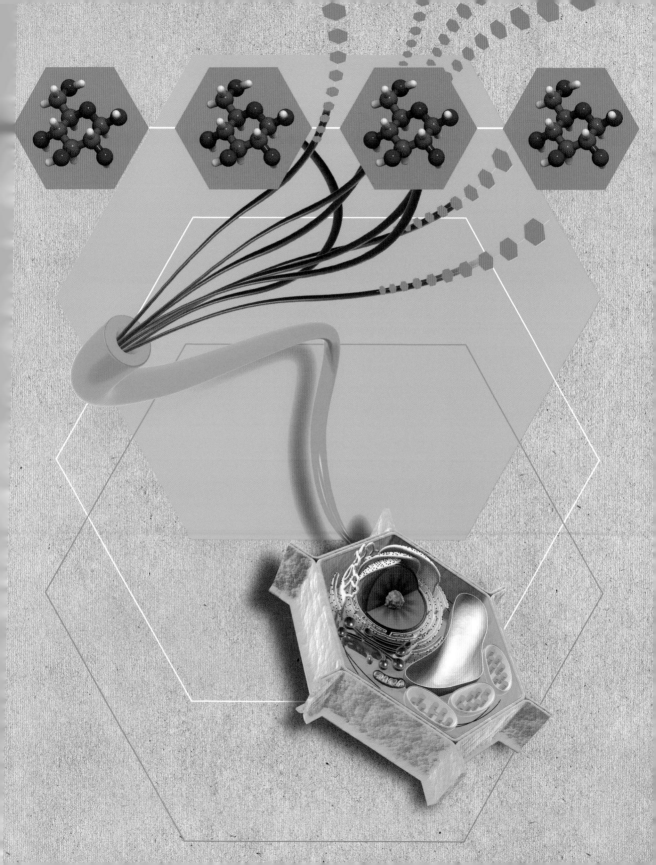

LIPIDS
the 30-second structure

The membranes that surround cells and the fat we use to store energy are comprised of lipids, molecules with enough greasy parts that enable them to dissolve in oily liquids but not water. This insolubility in water allows them to clump together in ways that depend on the lipid's shape and whether or not it has a part that is 'water-loving' or can stick tightly to water. The main lipid constituents of fat, triglycerides, lack water-loving parts and so clump together into oily droplets. These droplets serve as dense energy stores since each triglyceride contains three greasy, petroleum-like tails that can be burned to release energy. The structural lipids of biological membranes have two greasy 'tails' and one water-loving 'head' and are rod-like in shape. When placed in water, the rods stack in a wall-like sheet formation in which the water-loving heads all face one way and the greasy tails another. To keep the tails out of water, two such sheets stack to form a 'bilayer' in which the tail sides of the sheets face one another, giving a thin water-impermeable 'tail' layer capped by the water-loving heads that face outside. Signalling lipids are more diverse in structure. All work by binding to proteins to trigger a cascade of chemical reactions, changing what the receiving cell is doing.

RELATED TOPICS
See also
LIFE'S ORGANIZING FORCE
page 28

MEMBRANES, MESSENGERS
& RECEPTORS
page 66

LONG-TERM STORAGE
& BURNING FATS
page 80

BIOCHEMISTRY ON THE BRAIN
page 124

3-SECOND BIOGRAPHIES
OTTO WALLACH
1847–1931
German chemist who developed methods for determining the structures of plant lipids called terpenes

CHARLES ERNEST OVERTON
1865–1933
British biologist who first proposed that the membranes which surround cells are comprised of lipids

30-SECOND TEXT
Stephen Contakes

3-SECOND NUCLEUS
Lipids are greasy, clump together in water and release a lot of energy when burned. Structurally diverse, they serve as dense metabolic fuel reserves, membrane partitions and intercellular signals.

3-MINUTE SYSTEM
Some signalling lipids contain one or more rings of carbon atoms, from which dangle hydrocarbon parts and oxygen-containing parts. For instance, the prostaglandin hormones involved in triggering the body's response to infection and injury contain a five-membered ring made by linking two carbons from a fatty acid together. The drugs ibuprofen and aspirin block this ring-forming process and consequently the pain and inflammation that prostaglandins trigger.

Lipids stick together and form a bilayer in the cell membrane.

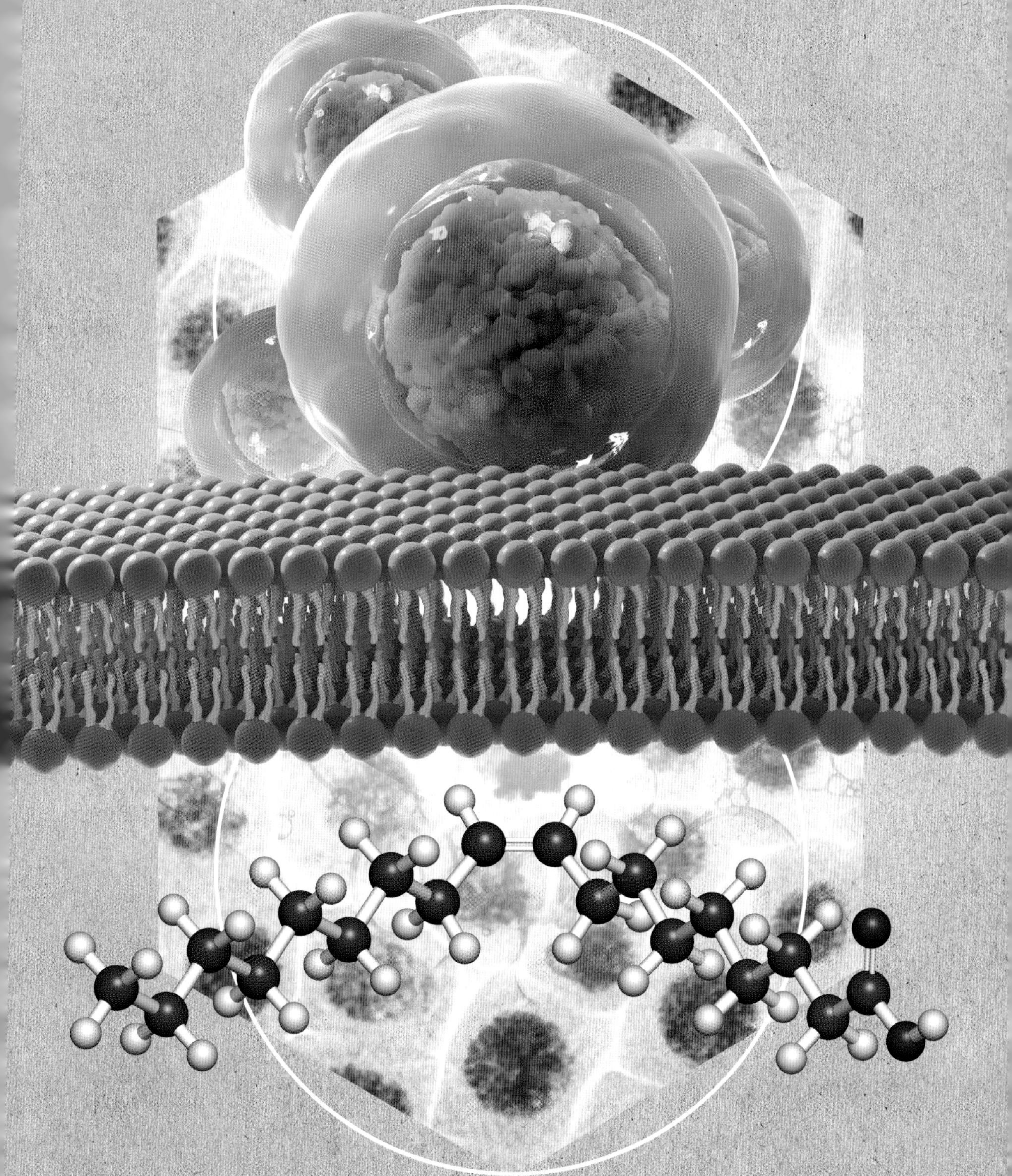

SWEETNESS: CARBOHYDRATES

the 30-second structure

3-SECOND NUCLEUS
Carbohydrates contain carbon rings which can be linked into fibrous and branched structures that act as metabolic energy reserves, fibres and information-bearing cellular recognition elements.

3-MINUTE SYSTEM
Carbohydrates on cell surfaces act as a kind of signal. For instance, on the surface of your red blood cells are a set of four or five linked monosaccharide rings that determine your blood type. The arrangement of –OH and other units protruding from the ring carbons defines a unique pattern of water-loving and greasy patches that fit into and stick to proteins outside the cell, much as a hand fits in a glove.

So named because chemically they are multiples of carbon ('carbo') and water (H_2O, 'hydrate'), carbohydrates are life's utility molecules. They include single-sugar units called monosaccharides, each consisting of a chain of three to seven carbon atoms, one of which is bound to a single oxygen (O) while the remaining carbon atoms are bound to one hydroxyl (–OH) each. The variable length of the chain, location of the unique oxygen and three-dimensional arrangement about some of the carbons gives rise to numerous monosaccharides, which may be split, rearranged and hooked together by exploiting the reactivity of the hydroxyl group. This chemical and structural flexibility enables a single carbohydrate to perform many metabolic functions. The most prominent, glucose, is formed in photosynthesis and serves as a fuel, releasing energy when it is broken down and capable of either being hooked together into polymers called polysaccharides or converted into fats for storage when it is not needed. Because one sequence of reactions for breaking down glucose gives the monosaccharide ribose, it can even be used to make nucleic acids and energy storage molecules such as ATP. Glucose even fulfils a structural role in plants and bacteria. Plants string it together into tough fibres, while some bacteria chemically modify it to build the outer walls of their cells.

RELATED TOPICS
See also
USING BIG CARBS
page 76

USING SMALL CARBS
page 78

BURNING IN REVERSE:
PHOTOSYNTHESIS
page 94

3-SECOND BIOGRAPHIES
EMIL HERMANN FISCHER
1852–1919
German chemist who helped elucidate carbohydrate's three-dimensional shapes

NORMAN HAWORTH
1883–1950
British chemist who determined the structures of carbohydrate rings and many di- and polysaccharides

30-SECOND TEXT
Stephen Contakes

Sugars: glucose (top) and fructose (bottom) are monosaccharides. Sucrose (table sugar, centre) is a disaccharide.

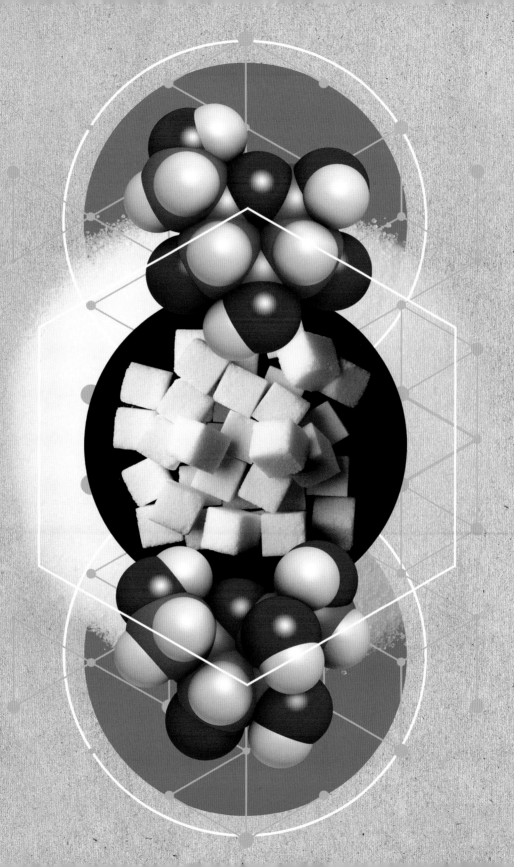

13 August 1919
Born in Rendcomb in Gloucestershire, England, where his father was the local GP

1932–5
Undertakes extensive laboratory work at Dorset's Bryanston School, stimulating his desire to pursue a life in science

1940
Marries fellow pacifist Joan Margaret Howe, receives a first-class degree in biochemistry from the University of Cambridge, and as a conscientious objector works as an orderly at Winford Hospital near Bristol

1943
Completes his PhD for work on the metabolism of the amino acid lysine and begins investigating the amino acid composition of proteins

1945
Develops a method for determining the number of polypeptide chains in a protein, now known as end-group analysis

1947–53
Successfully sequences insulin by breaking the chains into smaller fragments using enzymes, determining their composition and figuring out how the pieces fit together

1958
Awarded the Nobel Prize in Chemistry for his work on insulin

1977
Develops the easy-to-implement 'Sanger method' of DNA sequencing, which continues to find wide use today

1980
Shares the Nobel prize with Walter Gilbert and Paul Berg for his work on the sequencing of nucleic acids

1983
Retires from active scientific work to build boats and spend time gardening

19 November 2013
Dies in Cambridge, England

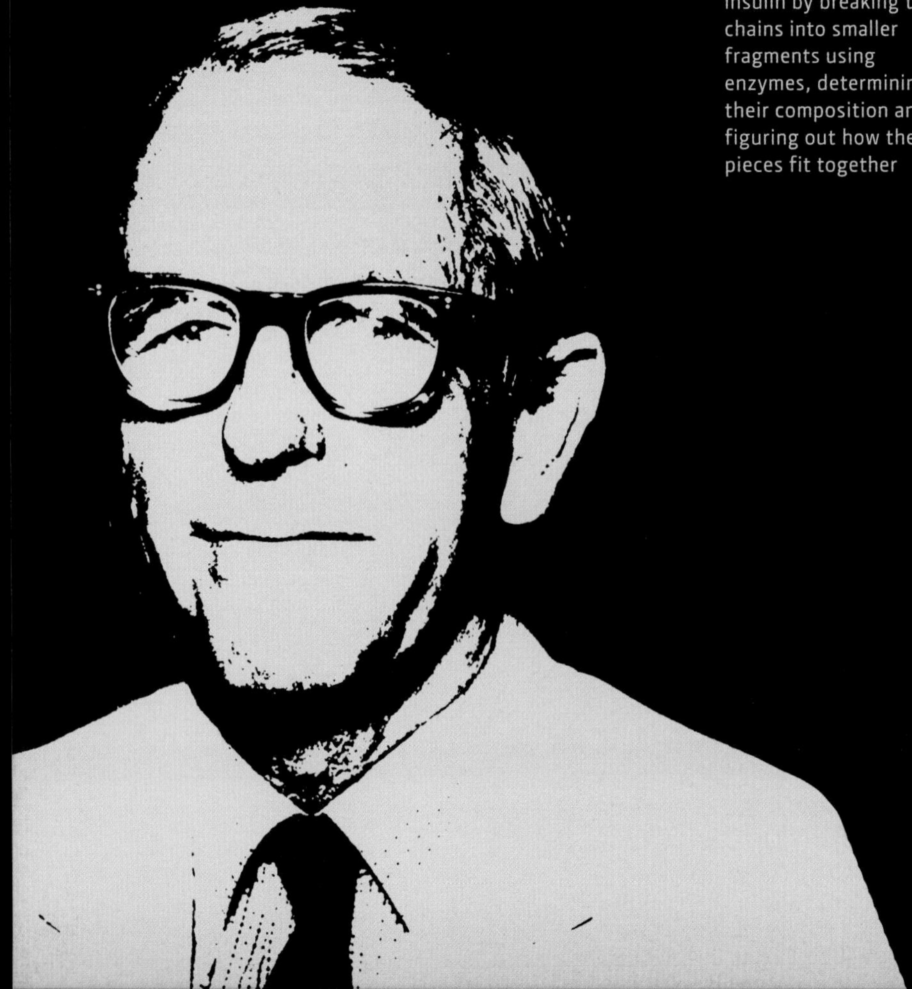

FREDERICK SANGER

Today it is taken for granted that scientists can determine the sequence of bases in an organism's DNA. Such sequences are routinely employed in fundamental research, medicine, agriculture, legal proceedings and genealogy research while, as of early 2021, the complete DNA sequences of more than 10,000 organisms are publicly available online. One of the scientists who contributed most significantly to these achievements is Fred Sanger, who developed ingenious methods for reading the chemical sequences of proteins and nucleic acids. He won two Nobel Prizes in Chemistry for his work.

Sanger was born to a wealthy GP in Gloucestershire in 1919. His interest in science was kindled during his studies at Bryanston School, after which he went on to study biochemistry at Cambridge. At the time he completed his PhD in 1943, proteins were erroneously suspected to contain organisms' genetic information. However, little was known about their structures, even though several had been isolated in pure form. These included the regulatory protein insulin, which was widely used to treat diabetes and could be purchased in pharmacies. This provided an occasion for Sanger to study its structure, which he did by attaching a yellow tag (now called Sanger's reagent) to one end of the protein chains. By separating the tagged chains Sanger was able to determine how many chains were present. He found that each molecule of insulin contains two polypeptide chains. Later, by selectively breaking apart the chains in various ways and studying the amino acids of which they were composed, he was also able to puzzle out the complete sequence of amino acids in each chain. For this he was awarded the 1958 Nobel Prize in Chemistry.

In the 1960s and 1970s Sanger extended his sequencing work to include RNA and DNA. He sequenced DNA using tiny amounts of didexoy nucleotides, which stop the growth of its chains and give off different colours for different bases (As, Gs, Cs and Ts). When DNA is copied in the presence of these nucleotides a mixture of chains of every possible length is formed. The DNA sequence can then be read from this mixture by looking at the light emitted by the chain-terminating dideoxy nucleotides (different in colour for As, Ts, Gs and Cs) as the fragments are separate in order from smallest to largest. This technique, now called Sanger sequencing, earned Sanger a share of the 1980 Chemistry Nobel. It is still widely employed today.

Stephen Contakes

LIFE'S LETTERS: NUCLEOTIDES

the 30-second structure

Nucleotides consist of a

phosphate attached to a five-membered ring derived from the carbohydrate ribose that in turn is linked to a flat nitrogen-containing structure called a nucleobase. There are four different nucleobases among the nucleotides that are used to encode and transmit genetic information. All have greasy flat faces. They differ from one another in the distinctive pattern of partially charged nitrogen, oxygen and hydrogen around their thin edges. When the riboses are linked through the phosphates to form chains of deoxyribonucleic acid (DNA), the flat bases stack together, causing the chain to twist into a helix. This produces a pattern of base edges characteristic of the sequence, in which the bases occur along the chain. This sequence is the genetic message. It can be read as bases with complementary edges selectively interact or base pair with the original bases to generate a sequence complementary to the first. When these bases are part of nucleotides containing deoxyribose, the result is a double helix of two complementary strands. DNA strands may be transcribed into a message in the form of polymers of ribose-based nucleotides called ribonucleic acids (mRNA). These translate the message (m) by directing other RNA molecules (tRNA) to line up in the right sequence and transfer (t) amino acids to a protein chain.

3-SECOND NUCLEUS
Nucleotides serve as energy carriers and form nucleic acid polymers, including those that encode the sequences of proteins.

3-MINUTE SYSTEM
Nucleotides also serve as carriers of energy in metabolism. The best known is adenosine triphosphate or ATP, which raises the energy of other biomolecules by donating one of its phosphate groups to them, leaving adenosine diphosphate (ADP) behind. This ADP can later receive a phosphate from high-energy, phosphate-bearing molecules to form ATP. The nicotinamide adenosine dinucleotide, NAD+, is equally important; it receives high-energy electrons released in biological oxidations and uses them to make ATP.

RELATED TOPICS
See also
LIFE'S CURRENCY
page 74

THE MAKING OF THE
MESSAGE BEARERS
page 100

MAINTAINING THE CODE
page 102

READING THE BOOK
page 106

3-SECOND BIOGRAPHIES
OSWALD THEODORE AVERY JR.
1877–1955
American biochemist who discovered that genes are made from DNA

ERWIN CHARGAFF
1905–2002
Austro-Hungarian-born biochemist who helped establish the existence of DNA base pairs

30-SECOND TEXT
Stephen Contakes

DNA is a polymer of deoxyribonucleotides that forms a double helix.

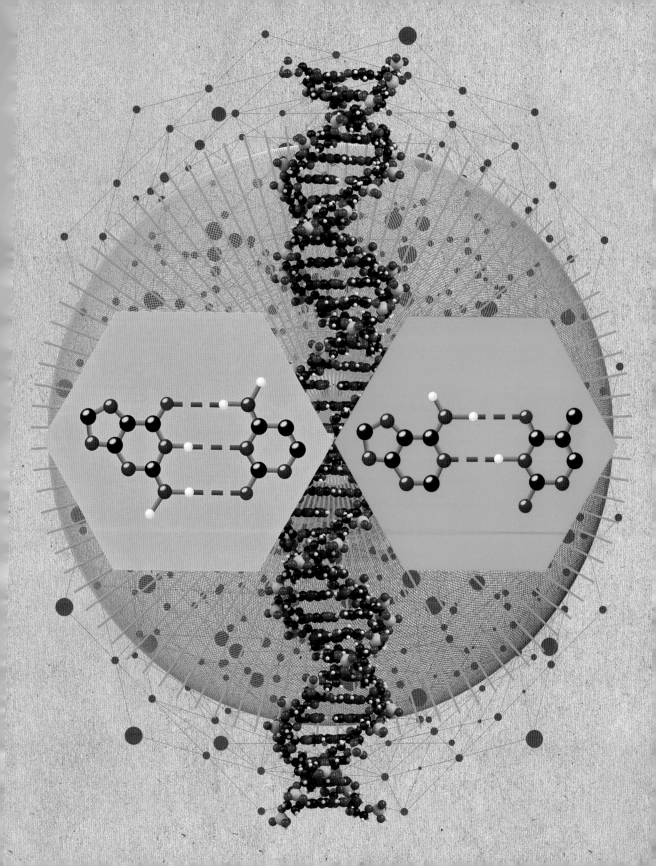

WORKFORCE: AMINO ACIDS & PEPTIDES

the 30-second structure

3-SECOND NUCLEUS
Amino acids are linked together to form protein chains, which carry out a staggering variety of biochemical functions.

3-MINUTE SYSTEM
The peptide hormone insulin promotes glucose storage when our bodies are in a well-fed state. Our body breaks down this glucose to obtain energy during times of metabolic need, such as when we are fasting or doing intense exercise. In type 1 diabetes, the body produces little or no insulin. Individuals with type 1 diabetes require daily injections of insulin to survive. Individuals with type 2 diabetes have cells that don't respond normally to insulin the body produces.

The vast array of proteins in our bodies, which have unique shapes and essential biochemical functions, are composed of 20 different amino acid building blocks. Amino acids are named for the amine (NH_2-) and carboxylic acid ($-CO_2H$) groups of atoms that are bonded to a central carbon atom. The central carbon atom also contains a variable group called a side chain that gives the amino acid its identity. The side chain can be acidic, basic or neutral; hydrophobic (greasy/insoluble in water) or hydrophilic (sticks well to water); large or small. Amino acids may be linked together to form chains in a process involving the loss of a water molecule. The resulting linkage between the amino acids is known as a peptide bond. Short chains of amino acids are referred to as peptides, whereas longer chains containing 50 or more amino acids are referred to as polypeptides or proteins. While proteins perform a variety of functions and are the main workforce of the cell, many of the peptides in our bodies act as hormones – chemicals that are secreted in one location and induce a physiological response elsewhere. Examples include insulin and glucagon, which work together to balance blood-sugar levels, and oxytocin, which plays an important role in reproduction and mother–infant bonding.

RELATED TOPICS
See also
BONDS: LIFE'S HANDEDNESS
page 36

CHAINS: BIOLOGICAL POLYMERS
page 38

PROTEIN FOLDING & AGGREGATION
page 54

SHAPE: PROTEIN STRUCTURE
page 56

3-SECOND BIOGRAPHIES
GERARDUS JOHANNES MULDER
1802–80
Dutch chemist who first described the composition of proteins

VINCENT DU VIGNEAUD
1901–78
Nobel Prize-winning American biochemist who performed the first synthesis of the peptide hormone oxytocin

30-SECOND TEXT
Kristi Lazar Cantrell

Amino acids form chains of peptides (such as insulin), which are vital for biological function.

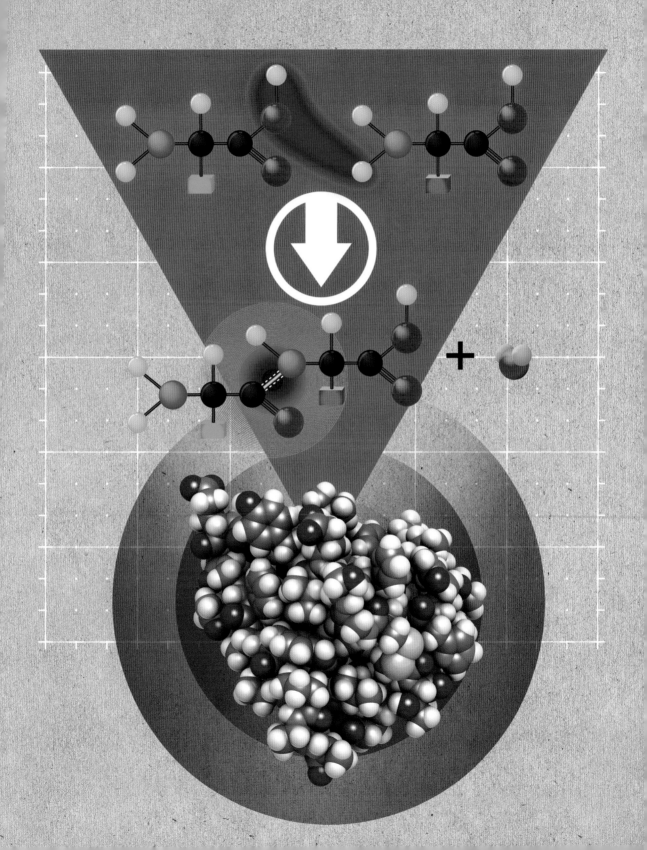

MACHINERY ◖

MACHINERY
GLOSSARY

adenosine triphosphate (ATP) A nucleotide that serves as an energy currency, being formed in processes that release energy and broken down in processes that use energy.

aggregate A collection of molecules held together through intermolecular attractions.

amyloid fibril A water-insoluble aggregate of misfolded proteins that forms a fibre-like structure. Deposits of amyloid fibrils are associated with conditions such as Alzheimer's, Huntington's and motor neurone disease.

cascade A series of chemical reactions that take place in response to a cellular signal, in which the signal is amplified as enzymes are activated and go on to catalyse many chemical reactions.

collagen A structural protein consisting of three protein chains coiled together into a rod-shaped triple helix. Fibrous aggregates of collagen triple helices are the most abundant form of protein in the human body.

cytoskeleton The network of fibrous proteins that acts as a skeleton for the cell by stabilizing it and giving it its shape.

enzyme A protein that acts as a catalyst by facilitating a chemical reaction.

ferritin An iron-storage protein consisting of a hollow, roughly spherical shell of protein encaging thousands of iron atoms present in the form of an iron oxide mineral.

globular protein A protein with a compact or globular shape; as opposed to fibrous proteins, which possess elongated shapes.

haemoglobin A globular protein found in the red blood cells of vertebrates that binds oxygen in the lungs and transports it around their bodies.

hydrophobic Something that appears to repel water since it doesn't stick tightly to it.

ion An atom or grouping of atoms possessing a positive or negative charge.

microtubule Tube-shaped aggregates of the globular protein tubulin that help comprise the cytoskeleton and act as roads for cargo-transporting motor proteins.

monomer Commonly refers to molecules that can react with one another to form longer molecular chains called polymers; in protein chemistry it also refers to individual proteins that associate to form larger aggregates.

motor proteins A class of proteins that converts chemical energy into mechanical energy.

myosin proteins A class of motor proteins involved in muscle contraction and the transport of cellular cargo.

nucleic acids DNA and RNA polymers of nucleotides. These are called nucleic acids because they are derivatives of phosphoric acid mainly residing in and around the cell nucleus; they are involved in the storage and expression of genetic information.

polypeptide A polymer chain formed by hooking amino acids together.

ribosome A collection of associated RNA and protein that helps link amino acids into proteins.

synthesis A chemical reaction that makes a desired product. In biochemistry it usually refers to the formation of larger or more complex molecules from smaller ones.

substrate A molecule that binds to the active site of an enzyme, where the enzyme facilitates its chemical transformation.

transferrin A globular protein present in blood plasma that delivers iron to cells.

PROTEIN FOLDING & AGGREGATION

the 30-second structure

3-SECOND NUCLEUS
Proteins possess a functional, folded three-dimensional structure called the native state. This can misfold, resulting in protein aggregation and disease.

3-MINUTE SYSTEM
Ribosomes, found within all living cells, are complex machines that perform protein synthesis. They link amino acids together to form nascent protein chains. As these chains emerge from the ribosome they are prevented from misfolding by a ribosome-associated protein chaperone called Trigger Factor. Trigger Factor, along with other chaperones it recruits to assist in the folding process, works to fold proteins into their native states.

The native state of a protein is its functional, folded three-dimensional structure. Proteins adopt their native state as the amino acid chains twist and turn relative to one another. This occurs in directed pathways where the energy is lowered as small local elements of structure form and then coalesce into larger structures. This protein-folding process is accompanied by hydrophobic collapse, in which water is expelled from around greasy hydrophobic amino acids that become buried inside the protein interior, while polar amino acids that can interact strongly with water are left to face towards the water-rich environment of the cell. In cells this process is monitored and controlled by hundreds of chaperone proteins. Chaperones jump in and protect misfolded proteins before they can aggregate with other nearby proteins and cause disease. The proteins misfold and aggregate when hydrophobic amino acids are exposed on protein surfaces and stick together with the hydrophobic parts of other proteins. Aggregation can result in a range of human diseases. The aggregates may be unstructured amorphous aggregates or highly ordered aggregates referred to as amyloid fibrils. Many human diseases are associated with amyloid fibril formation, including the neurodegenerative diseases Alzheimer's, Huntington's and Parkinson's disease.

RELATED TOPICS
See also
LIFE'S ORGANIZING FORCE
page 28

WORKFORCE: AMINO ACIDS & PEPTIDES
page 48

SHAPE: PROTEIN STRUCTURE
page 56

READING THE BOOK
page 106

3-SECOND BIOGRAPHIES
CHRISTIAN ANFINSEN
1916–95
American biochemist who showed that the information needed for a protein to adopt its native three-dimensional structure is encoded within the amino acid sequence

CYRUS LEVINTHAL
1922–90
American molecular biologist who showed that protein folding is not random

30-SECOND TEXT
Kristi Lazar Cantrell

Proteins fold into their native structure. Misfolded proteins (right) result in disease.

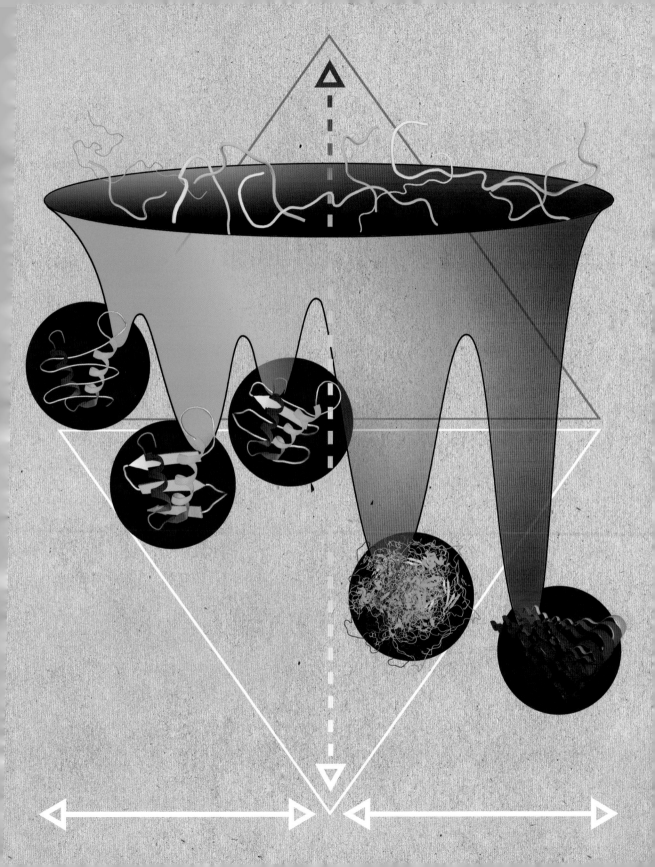

SHAPE: PROTEIN STRUCTURE

the 30-second structure

Every protein contains a unique amino acid sequence that determines the three-dimensional structure it will adopt. Forming the correct structure is important because many proteins rely on the recognition of specific molecular shapes to function correctly. The polypeptide chain folds into a three-dimensional structure as the side chains of the amino acids interact with each other and the surrounding water. Many proteins adopt a compact, globular structure that is soluble in water. Globular proteins contain two stable structural elements that arise from interactions between amino acids: screw-like alpha-helices and beta-sheets that resemble curtains due to their pleated edge-on appearance. For example, the digestive enzyme alpha amylase is rich in both alpha-helices and beta-sheets. Its overall structure enables it to bind to carbohydrates in saliva and break them apart. In contrast, most amino acids in antibody proteins form beta-sheets. Between two of the sheets is a site that binds foreign invaders, such as viruses, enabling antibodies to target them for destruction. Polypeptide chains can also clump together in various ways. Found in red blood cells, haemoglobin is an aggregate of four chains with alpha-helical structure, each of which can bind oxygen. This structure helps it effectively transport oxygen from the lungs to tissues.

3-SECOND NUCLEUS
Every protein has a unique amino acid sequence that dictates the shape of the protein and its biochemical function.

3-MINUTE SYSTEM
Some proteins are fibrous, forming stiff and elongated fibres. They are insoluble in water and provide structural support in our bodies. Fibrous proteins include alpha-keratin in hair and fingernails, fibroin in silk and spider webs, and collagen in bones, muscles, skin and tendons. Collagen is the most abundant protein in the human body. Fibrous proteins tend to have repetitive amino acid sequences, in contrast to globular proteins, which have non-repetitive sequences.

RELATED TOPICS
See also
PROTEIN FOLDING
& AGGREGATION
page 54

MAX PERUTZ
page 58

COORDINATING THE DEFENCE
page 130

3-SECOND BIOGRAPHIES
LINUS PAULING
1901–94
American chemist who showed the importance of the alpha-helix and beta-sheet in protein structure

JOHN KENDREW
1917–97
British biochemist who, with Max Perutz, won the Nobel Prize for studies on the structures of haemoglobin and myoglobin

30-SECOND TEXT
Kristi Lazar Cantrell

Beta sheets and alpha helices are structural elements of proteins such as alpha-amylase and haemoglobin.

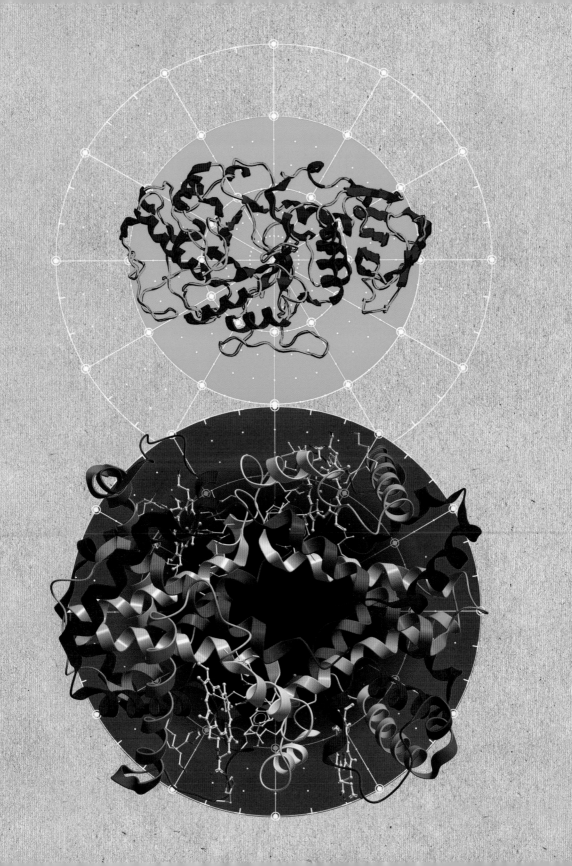

19 May 1914
Born in Vienna, Austria

1936
Completes undergraduate work in chemistry at the University of Vienna and begins graduate work at Cambridge

1938
With the glaciologist Gerald Seligman explains how glaciers flow

1940
Completes his PhD under the pioneering crystallographer Lawrence H. Bragg

1947
Establishes the Medical Research Council Unit for study of the molecular structure of biological systems. Starting with just Bragg and John Kendrew, work in this unit will eventually produce 12 Nobel Prizes

1954
Working with David Green and Vernon Ingram Perutz develops the heavy atom method for solving the 'phase problem' in crystallography, permitting the structures of molecules to be determined directly using X-ray crystallography

1959
Uses crystallography to determine the three-dimensional structure of the globular protein haemoglobin

1962
Shares the Nobel Prize in Chemistry with John Kendrew, who had determined the structure of the oxygen storage protein myoglobin

1970
Perutz's continued work on haemoglobin explains in detail how haemoglobin binds oxygen tightly in the lungs and releases it in muscle

1979
Retires from laboratory administration but continues to perform research and writes for the popular press, through which he acquires a reputation as a science communicator

6 February 2002
Dies in Cambridge, England

MAX PERUTZ

When you take a breath, the oxygen you draw in sticks to the protein haemoglobin, which transports it through the bloodstream to your muscles, where it then hands it off to another protein called myoglobin. To do this well, haemoglobin must grasp molecules of oxygen tightly in the lungs but hold them more loosely in muscle. How it does this was a mystery before the work of Max Perutz, who first determined haemoglobin's structure and demonstrated why it works the way it does.

Born in 1914 to a prosperous textile mill owner, Perutz's interest in biochemistry was awakened at the University of Vienna. He began his work on proteins under the Irish crystallographer John Desmond Bernal at the University of Cambridge, where he would spend most of his career. Under Bernal he learned X-ray crystallography, which uses the pattern of spots generated when X-rays are passed through a crystal to determine its internal structure. At that time, crystallography had only been applied to simple structures as proteins were considered far too complex. Nevertheless, after being first tasked with studying a silicate mineral Perutz obtained a 'diffraction pattern' for haemoglobin. Encouraged by this success, his PhD mentor

Lawrence Bragg (himself a Nobelist and pioneering structural biologist) arranged for Perutz to be appointed head of the laboratory for study of the molecular structure of biological systems at Cambridge, which then consisted only of himself and John Kendrew.

Despite Perutz's success in obtaining a diffraction pattern for haemoglobin, it was not clear how the pattern could be translated into haemoglobin's structure. For that Perutz needed to know the state or 'phase' of the X-rays as they interacted with each atom, which nobody knew how to figure out at that time. After six years of work Perutz solved this problem using mercury atoms, which he dangled off haemoglobin in a way that changed its diffraction pattern but not its overall structure. This enabled Perutz to work out the phases of the X-rays and solve haemoglobin's three-dimensional structure. For this Perutz shared the 1962 Nobel Prize in Chemistry with John Kendrew, who had used Perutz's technique to find the structure of myoglobin.

Later Perutz determined the structures of haemoglobin both with and without oxygen in great detail and was able to explain how it works. Afterwards, he retired from laboratory administration, though he continued to perform research until shortly before his 2002 death.

Stephen Contakes

ROADS & RAFTERS: STRUCTURAL PROTEINS

the 30-second structure

3-SECOND NUCLEUS
Structural proteins
compose flowing protein
skeletons for cells, paths
for other proteins to move
along and fibres outside
cells such as tendons
and hair.

3-MINUTE SYSTEM
The death cap mushroom
contains phalloidin, a
deadly molecule that
binds actin. Many poisons
break things apart, but
this poison holds things
together. Phalloidin
stabilizes actin domains so
that they never leave the
fibre, giving the fibre too
much stability and
gumming up the inner
workings of the cell. This
poison has proved useful
in the lab: a fluorescent
version of phalloidin binds
actin fibres and makes even
individual fibres glow under
a microscope.

Up to 5 per cent of the protein
in a typical cell is a tangled web of fibres called
actin. These fibres provide a skeleton for the
cell, lending strength, structure and shape. This
cytoskeleton is much more dynamic than bone,
constantly changing according to the needs of
the cell. In actin fibres, individual actin monomer
proteins bind end-to-end, like pavement stones
in a row. Over time, an actin monomer at one
end destabilizes and falls off the fibre, while
another can bind to the other end, maintaining
the same number of actin domains overall in a
state of dynamic instability. Actin also provides
a mechanism for motions within a cell – myosin
proteins grab on to actin domains in the actin
fibre, which act like a ladder for the myosin
proteins to climb up. Some myosin proteins
carry molecular and organelle cargo around the
cell; others are responsible for cellular motions.
Because of this, muscle cells contain higher
levels of actin – as much as 20 per cent in some
cases – supporting larger movements such as
muscle contraction. Outside cells, structural
proteins perform diverse jobs. The tendons that
connect muscles to bones are made of collagen,
which is made of tightly wound, extremely
strong protein fibres. Strands of hair are woven
from fibres of a different protein called keratin.

RELATED TOPICS
See also
LIFE'S UNITS
page 16

SHAPE: PROTEIN STRUCTURE
page 56

MOTORS
page 62

MOVING
page 120

3-SECOND BIOGRAPHIES
FEODOR LYNEN
1911–79
German biochemist who
isolated the poison phalloidin,
which binds actin with high
affinity

BRUNÓ FERENC STRAUB
1914–96
Hungarian biochemist who
discovered actin fibres

30-SECOND TEXT
Ben McFarland

*A diversity of structural
proteins maintain cell
structure and shape,
including (left to right)
actin and myosin,
collagen and keratin.*

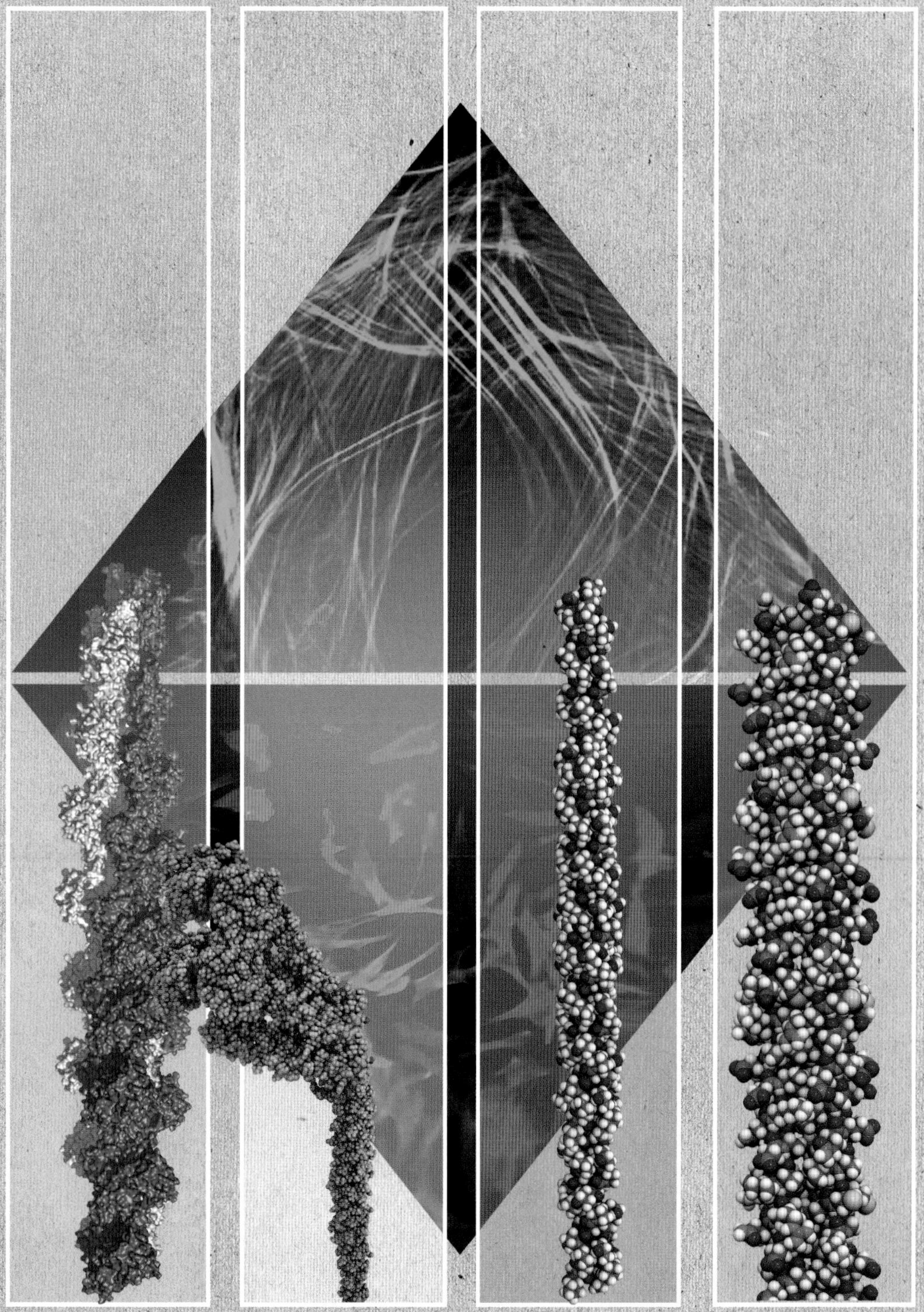

MOTORS

the 30-second structure

Just as the motor of a car uses the energy in petrol to turn its wheels, protein molecular motors use metabolic energy to drive the motions of living things. The motor proteins kinesin and dynein act as molecular pack horses, moving organelles and other cargo along microtubule roads. Other motors pump ions across membranes, drive the movement of enzymes along DNA chains or cause filamentous structures to move relative to one another. Taken in aggregate, the effect of the latter drives muscle contraction, enabling animals to run, crawl and swim. Yet all depend on metabolic energy, usually in the form of a molecule of ATP. When the motors kinesin, dynein and myosin grip a filament surface through their hand-like 'heads' a bound ATP is consumed, causing one of the heads to move past the other 'hand over hand' along the filament, just as a climber's hands move when scaling a rope. The rotary molecular pump vacuolar ATPase uses ATP to rotate a drum of membrane-bound proteins, in the process causing a proton (H^+ ion) to move across to the more positive side of the membrane. In contrast, the motors through which mitochondria power our cells use an imbalance of protons across a membrane to drive rotary motions that cause the synthesis of ATP.

3-SECOND NUCLEUS
Motor proteins convert the chemical energy in a molecule of ATP into motions that are used to contract muscles, cause hair-like structures on cell surfaces to beat and transport cargo around cells.

3-MINUTE SYSTEM
The wiggling of the tail-like structure of swimming sperm is driven by a host of dynein motors. The dynein motors are part of the tail's core, which contains two fixed microtubules running its length surrounded by a circle of nine doublet microtubules. Each doublet microtubule consist of two microtubules linked by a shared wall. One microtubule acts as a track; motors dangle off the sides of the other. The wiggling occurs when some doublets' motors march in step along the tracks of their neighbours.

RELATED TOPICS
See also
ROADS & RAFTERS: STRUCTURAL PROTEINS
page 60

ATP FROM AIR
page 86

MOVING
page 120

3-SECOND BIOGRAPHIES

VLADIMIR ENGELHARDT & MILITSA LYUBIMOVA
1894–1984 & 1898–1975
Soviet biochemists (husband and wife) who, in 1939, discovered that myosin splits ATP to produce muscle energy

IAN READ GIBBONS
1931–2018
British cell biologist who discovered dynein

RONALD D. VALE
1959–
American biochemist who discovered kinesin

30-SECOND TEXT
Stephen Contakes

Some motor proteins move biomolecule-rich sacs along microtubule roads.

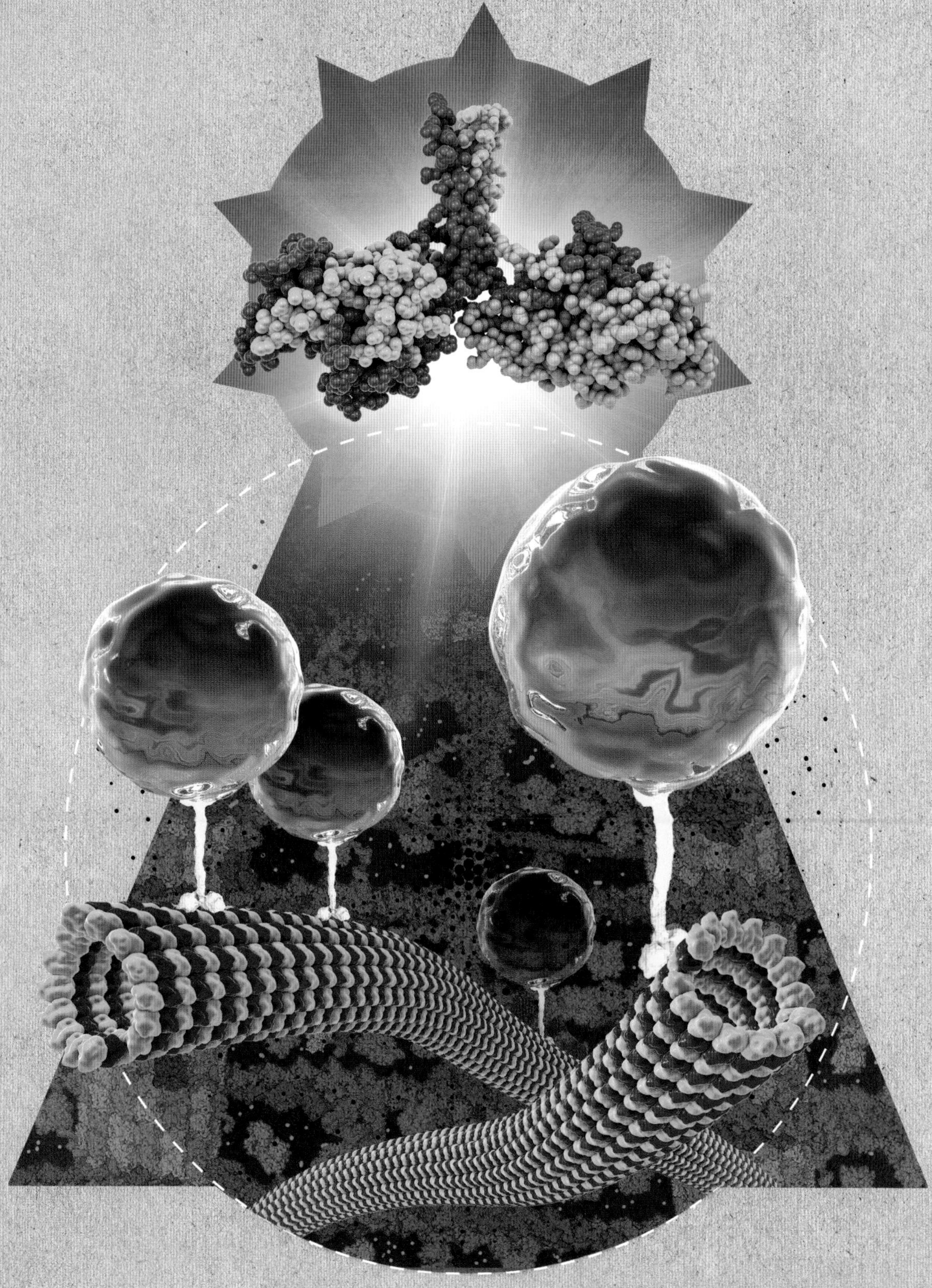

TRANSPORT & STORAGE

the 30-second structure

3-SECOND NUCLEUS
Some proteins bind small molecules to store them for later needs; others so they can transport them around.

3-MINUTE SYSTEM
More than half of the body's iron is in haemoglobin, which gives oxygenated blood its bright red colour. The oxygen is found deep inside, without an obvious path to the surface. So how did it get there? When dissolved in blood, haemoglobin's protein chains constantly ripple and move, temporarily opening up small crevices that oxygen can slip through to worm its way inside. This subtle motion is called, appropriately enough, breathing.

Proteins have chemical shapes that can fit together with other chemical shapes like three-dimensional jigsaw puzzles. Small molecules that fit together with and bind to proteins are called ligands. Storage proteins use this binding to act as boxes, holding ligands until they are needed. At any given time, about a quarter of the body's iron ions are stored in the liver in the form of ferritin. Ferritin is a hollow protein sphere that surrounds a nanoscale cluster of an iron mineral. In the blood, the iron is bound to transferrin, which dissolves in the blood and carries iron throughout the body. Some iron is transferred to the proteins haemoglobin and myoglobin, which use it to bind molecules of oxygen. Haemoglobin binds oxygen tightly in the lungs and transports it through the blood, releasing it in the tissues that need oxygen's chemical power. There it binds to the storage protein myoglobin, which holds it in reserve for when it can be used to burn fuels for energy. Some transport proteins are embedded in membranes. These move substances from one side of the membrane to the other, some by acting as pores and others by binding ligands on one side and releasing them on the other.

RELATED TOPICS
See also
MAX PERUTZ
page 58

BLOOD, BREATH & BALANCE
page 118

BIOCHEMISTRY ON THE BRAIN
page 124

ELECTROLYTE BALANCE
page 128

3-SECOND BIOGRAPHIES
VILÉM LAUFBERGER
1890–1986
Czech scientist who first isolated and crystallized ferritin

PETER AGRE & RODERICK MACKINNON
1949– & 1956–
American biochemists who won the 2003 Nobel Prize in Chemistry for helping establish how water and ions cross cell membranes

30-SECOND TEXT
Ben McFarland

Ferritin (top) is the main iron storage protein in the body. Transferrin (bottom) transports iron through the blood.

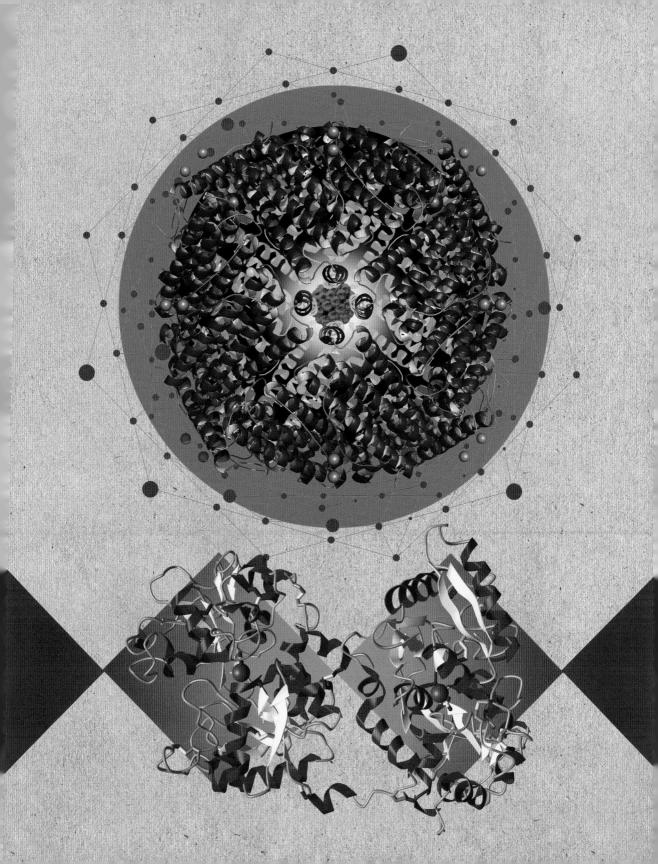

MEMBRANES, MESSENGERS & RECEPTORS

the 30-second structure

3-SECOND NUCLEUS
Receptor proteins
embedded in cell
membranes respond to
signals; those which
respond to smell, taste and
light form the biochemical
basis of our senses.

3-MINUTE SYSTEM
Opsins can respond to
light in odd places, such as
fish skin and fruit-fly ears.
In some types of animal
skin and hair, opsin
responds to cycles of light
and dark to help modulate
circadian rhythms. Some
cells in early invertebrates
contained light-responsive
opsins alongside taste
receptors, suggesting that
the first sensory cells could
both taste and see at the
same time.

The cell membrane forms a thick wall against the outside world, but sometimes the cell needs to respond to things outside. Protein receptors are like windows in the wall that let the cell sense its environment. In the presence of a signal some respond by causing metabolic changes on the inside of the membrane; others by opening to allow ions to flow across. A few work more slowly, providing the cell with new functionality as they trigger the synthesis of particular proteins. All five of our senses involve receptor-triggered processes. Those for vision, smell and certain tastes work similarly. When you smell something, a volatile molecule binds to the outside end of a receptor in your nose. When it does it shifts the receptor's shape inside the cell, sending another protein to be released. This protein, called a G protein since it binds a nucleic acid called a GTP nucleotide, starts a cascade of reactions, sometimes many millions strong. The light you see is absorbed by proteins called opsins. These contain a small non-protein group called rhodopsin, which changes its shape when it absorbs light. This in turn shifts the shape of the opsin and starts a signalling cascade so intense that some suggest the human eye can respond to even a single photon of light.

RELATED TOPICS
See also
MOBILIZING THE CELL
page 88

BIOCHEMISTRY ON THE BRAIN
page 124

PROGENY & THE PILL
page 140

3-SECOND BIOGRAPHIES
ROBERT LEFKOWITZ
& BRIAN KOBILKA
1943– & 1955–
American physiologists who
won the 2012 Nobel Prize in
Chemistry for discovering how
G protein-coupled receptors
work

RICHARD AXEL & LINDA BUCK
1946– & 1947–
American biologists who
won the 2004 Nobel Prize in
Physiology or Medicine by
mapping the molecular basis
of the sense of smell

30-SECOND TEXT
Ben McFarland

*A signalling cascade
is set off in the cell
membrane as G protein
coupled receptors
react to light.*

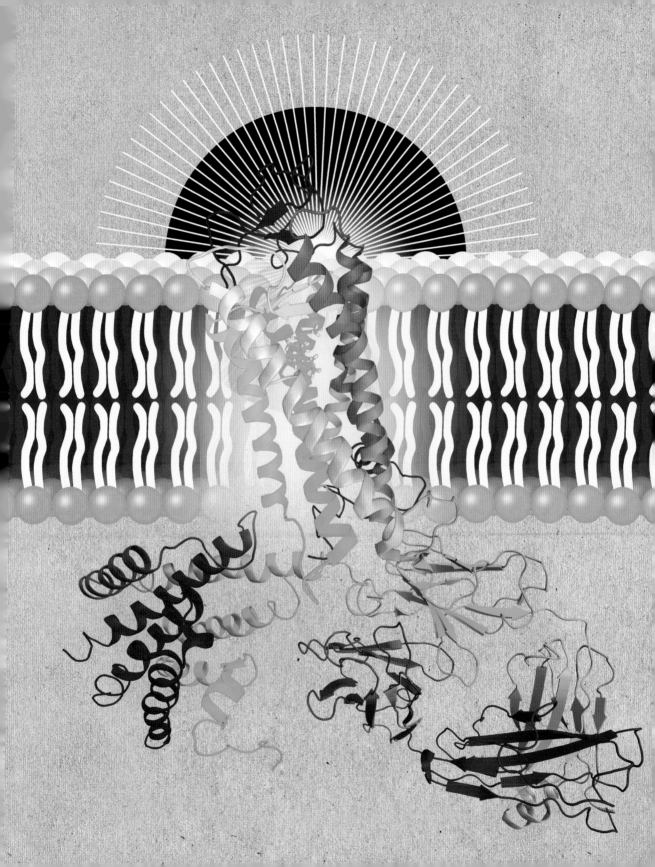

MAKING LIFE GO: ENZYMES

the 30-second structure

Chemical reactions need a minimum amount of energy to make them happen. This energy, called the activation energy, is typically used to break chemical bonds or contort molecules. The lower the activation energy of a reaction, the faster it occurs. Substances that lower the activation energy and make reactions go faster are called catalysts. Enzymes are proteins that act as biological catalysts, speeding up reactions in the body that otherwise might take too long to sustain life. Enzymes participate in all aspects of metabolism, such as digestion and the immune response. Enzymes lower the activation energy of a chemical reaction by interacting with the reacting molecules, called substrates. Each enzyme has an active site, which is where the reaction takes place. Depending on the function of the enzyme, these sites may adjust their shape to accommodate a range of substrates or else be highly selective, so that only certain substrate molecules fit, just as a key fits in a specific lock. Either way, once the substrates are bound in the active site, the chemical reaction takes place. The reaction products are released from the pocket, freeing the enzyme to start all over again. One enzyme can therefore catalyse the reaction many times over by continuously binding new substrates.

RELATED TOPICS
See also
SHAPE: PROTEIN STRUCTURE
page 56

LIFE'S CURRENCY
page 74

READING THE BOOK
page 106

DETOX
page 126

3-SECOND NUCLEUS
Hurry up! Proteins called enzymes speed up chemical reactions so that our metabolism can keep pace with the demands of life.

3-MINUTE SYSTEM
Many antibiotics work by blocking the ability of bacterial enzymes. Penicillin kills pathogenic bacteria by binding to the active sites of the enzymes they use to build their outer cell walls. This makes the enzyme unavailable for cell wall synthesis, which ultimately kills the bacteria.

3-SECOND BIOGRAPHIES
ANSELME PAYEN
1795–1871
French chemist who discovered the first enzyme, diastase

WILHELM KÜHNE
1837–1900
German physiologist who defined the term 'enzyme'

JAMES B. SUMNER
1887–1955
American chemist who purified the first enzyme, urease

30-SECOND TEXT
Andrew K. Udit

Enzymes provide a special site for reacting molecules to fit – like putting together puzzle pieces.

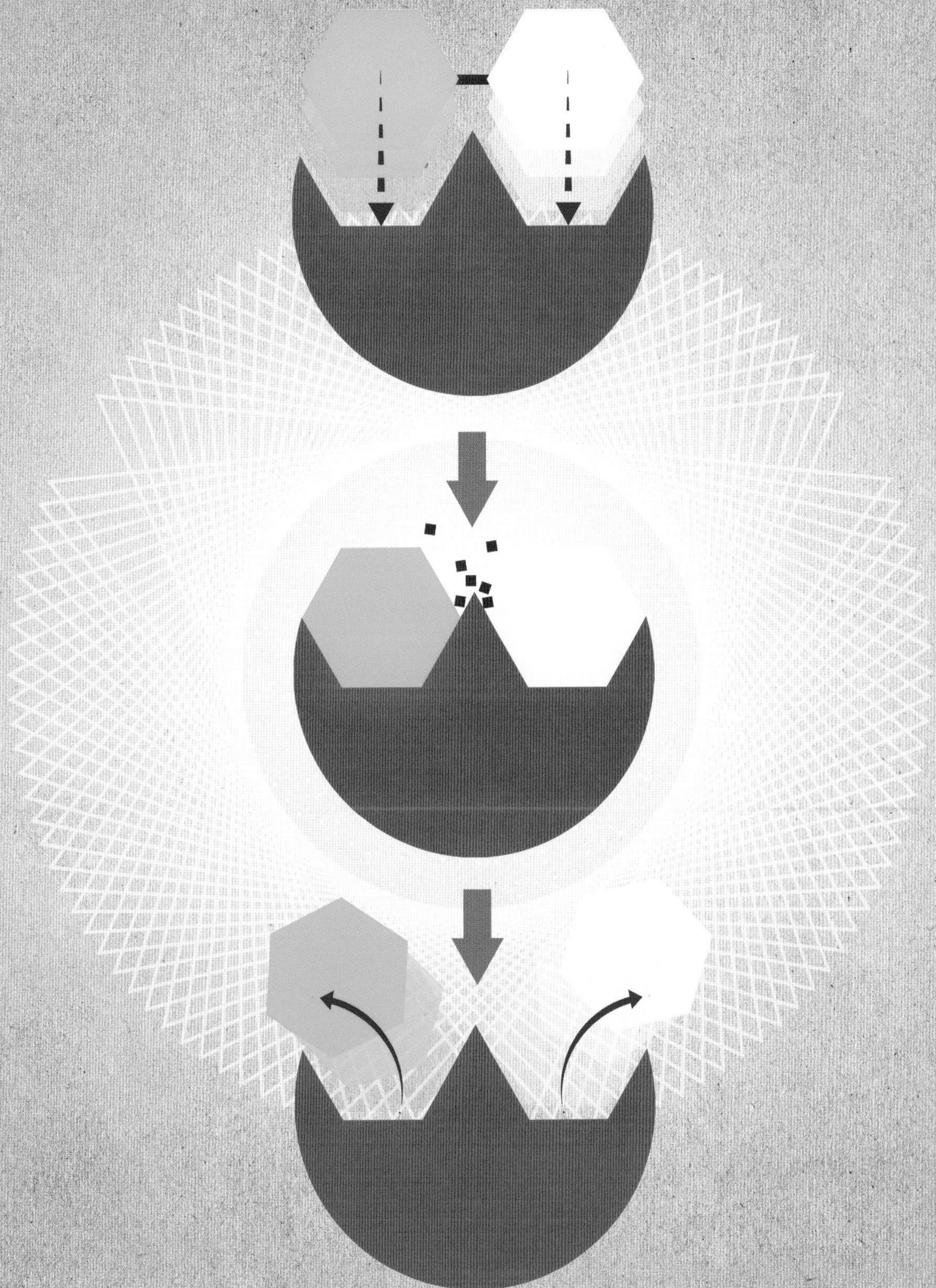

HARVESTING ENERGY

HARVESTING ENERGY
GLOSSARY

acetyl CoA Metabolic intermediate consisting of a two-carbon acetyl group attached to coenzyme A.

adipose tissue Another term for fat tissue, which serves as a long-term store of energy-rich triglycerides in mammals.

ADP Adenosine diphosphate, the molecule commonly formed when ATP is consumed.

adrenaline Hormone produced by the adrenal glands that triggers the rapid mobilization of metabolic energy (also called epinephrine).

aerobic Metabolic states and processes that work together to generate energy in processes that use oxygen, specifically via the breakdown of acetyl groups in the citric acid cycle and use of the resulting NADH and $FADH_2$ for ATP synthesis via the oxygen-requiring electron transport chain and oxidative phosphorylation.

anabolism The collection of metabolic reactions that uses energy to make larger, more reduced molecules from smaller, more oxidized ones.

anaerobic Metabolic states and processes that do not rely on oxygen.

beta oxidation The process by which fatty acids are broken down into acetyl CoA.

catabolism The collection of metabolic reactions that produces energy by breaking down larger, more reduced molecules into smaller, more oxidized ones.

Cori cycle A cycle that takes place during times of rapid energy use in which the liver supplies glucose to muscle, takes up the lactate produced in muscles and uses it to make more glucose.

endocrine system Physiological system that helps regulate metabolic processes at the organism-wide level by releasing hormones.

glucagon Hormone released by the pancreas that raises blood glucose levels by stimulating processes that produce energy.

gluconeogenesis The metabolic pathway by which glucose is resynthesized from pyruvate or lactate.

glycerol A three-carbon alcohol that can be linked with three fatty acids to give a triglyceride.

glycogen Bush-like polymer of glucose that is used to store glucose in muscles.

glycogenesis The metabolic pathway by which glycogen is synthesized.

glycogenolysis The metabolic pathway by which glucose units are released from glycogen.

glycolysis The metabolic pathway by which ATP and NADH are produced by converting glucose to pyruvate.

insulin A hormone released by the pancreas that serves to lower blood glucose levels by stimulating glucose uptake by muscle and adipose tissue and the biosynthesis of energy storage molecules such as glycogen.

lactate A three-carbon molecule made when glucose is broken down anaerobically.

lipogenesis The metabolic synthesis of lipids, typically triglycerides, starting from acetyl CoA.

lipolysis In a narrow sense refers to the metabolic breakdown of lipids, typically triglycerides, into glycerol and fatty acids, although depending on the context it can also include the breakdown of those fatty acids to give acetyl CoA.

metabolic pathway A set of linked metabolic reactions that together accomplish a particular metabolic function. For example, the glycolysis pathway comprises ten reactions that serve to produce two ATP and NADH while breaking down a glucose into two pyruvates.

mitochondria The power-plant organelles that produce most cellular ATP.

oxidation Chemical processes that remove electrons from atomic centres. In biochemistry this typically involves removing hydrogen atoms or increasing the proportion of oxygen atoms present in a molecule or ion.

proton A subatomic particle that chemists commonly abbreviate as H^+ since it may be formed by removing an electron from a typical hydrogen atom.

pyruvate A three-carbon metabolic intermediate produced from the breakdown of sugars.

reduction Chemical processes that add electrons to atomic centres. In biochemistry this typically involves adding hydrogen atoms or decreasing the proportion of oxygen atoms present in a molecule or ion.

LIFE'S CURRENCY

the 30-second structure

3-SECOND NUCLEUS
Living organisms use ATP
in their energy transactions
just like we use money:
working to earn it and
spending it to buy goods
and services.

Living organisms exchange mass
and energy with their surroundings all the time.
That is, they take in substances, they release
substances, they take in energy and they release
energy. This is the essence of life. When they
stop performing this trade, organisms die. The
exchange of mass and energy is made possible
through numerous intricate and intertwined
chemical reactions (that is, interactions between
substances – reactants – that result in the
formation of new substances – products).
The sum of these reactions is what we call
metabolism. Sunlight is the original source of
energy for plants; foods are the original source
of energy for animals, including us humans. Part
of this energy (about a quarter) is captured in
the synthesis of adenosine triphosphate, or ATP,
and stored in its chemical structure (the rest of
the energy is released as heat). We call ATP the
energy currency of living organisms because it
is the principal substance used in their energy
transactions. In a nutshell, organisms exploit
the energy from sunlight or foodstuffs to
synthesize ATP, and they spend energy by
breaking down ATP. Thus, they use ATP the
way we use money in everyday life – working
to earn it and spending it to meet our needs.

3-MINUTE SYSTEM
We divide metabolism into
catabolism, which includes
processes that degrade
large molecules to smaller
ones, and anabolism, which
includes processes that
synthesize large molecules
from smaller ones.
Catabolism yields raw
materials for anabolism
and energy (part of which
is stored in ATP) for
biological processes that
require energy such as
movement and anabolism.
In turn, anabolism serves
the development of an
organism, its maintenance
and the formation of
energy depots such as
starch, glycogen and fat.

RELATED TOPICS
See also
USING BIG CARBS
page 76

USING SMALL CARBS
page 78

LONG-TERM STORAGE
& BURNING FATS
page 80

THE CENTRAL HUB
page 82

3-SECOND BIOGRAPHIES
EDUARD BUCHNER
1860–1917
German chemist who, in
1897, discovered cell-free
metabolism – a landmark in
biochemistry and the start
of detailed metabolic studies

FRITZ LIPMANN
1899–1986
German-American biochemist
who, in 1941, proposed ATP is
central to energy exchange in
living organisms

30-SECOND TEXT
Vassilis Mougios

*The molecule ATP
regulates the way living
organisms use and
release energy.*

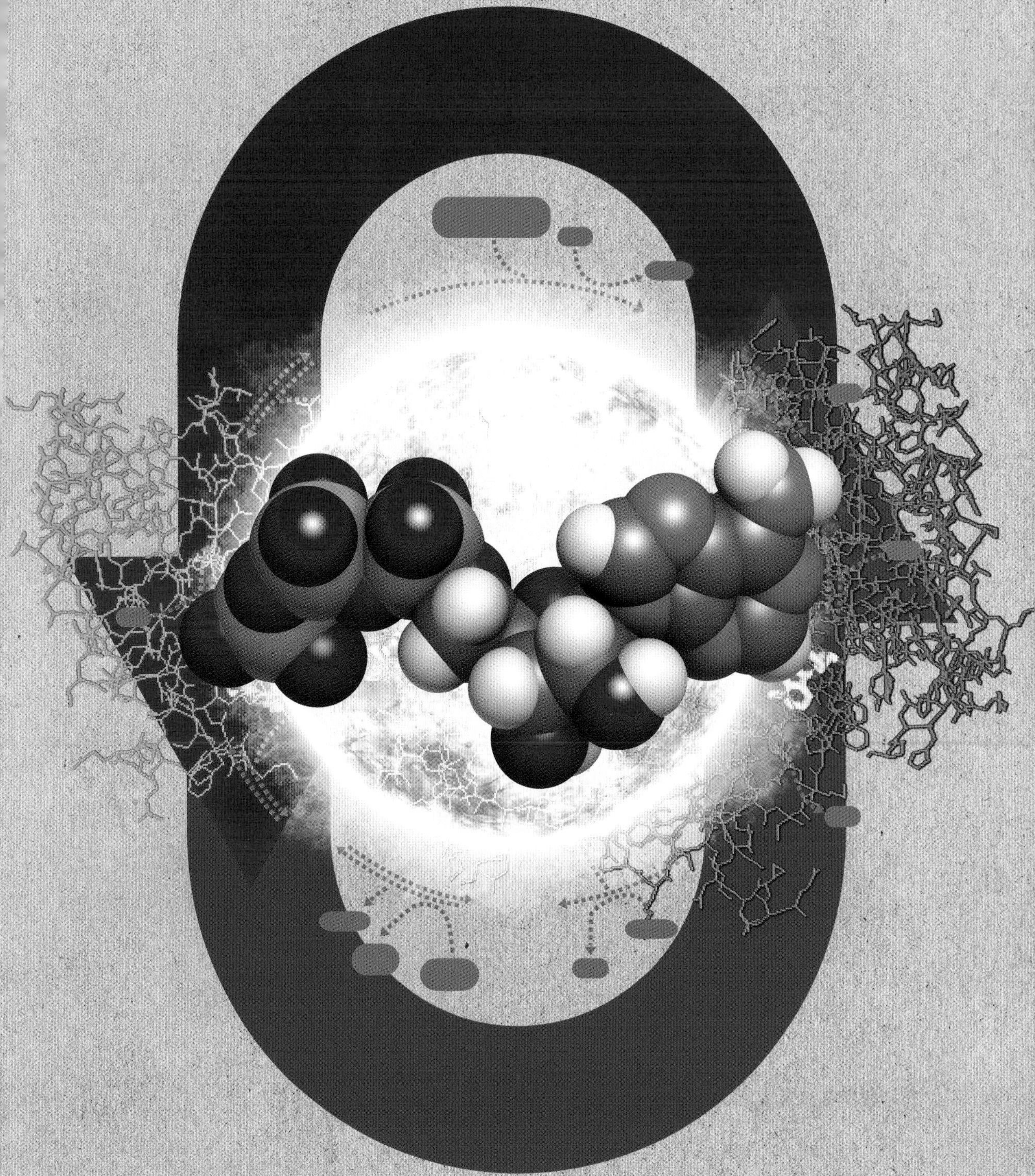

USING BIG CARBS

the 30-second structure

The biggest carbohydrate

molecules, called polysaccharides, comprise thousands of small units, the monosaccharides. Polysaccharides serve several purposes, one of them being energy storage. Plants and animals possess similar polysaccharides: starch and glycogen, respectively. Starch fills the seeds of plants and offers energy to the developing seedling until it launches its leaves to harvest energy from the sun. When animals (humans included) eat starchy foods, such as bread, cereals, pasta, potatoes, pulses and nuts, they digest starch in their gastrointestinal tract, producing glucose (the monosaccharide making up starch). Glucose is then transported by the blood to nourish all cells in the body. The liver and muscles store excess glucose in the form of glycogen by stringing the glucose molecules together in a process termed glycogenesis. When, several hours after a meal or during fasting, the blood glucose concentration drops, the liver breaks down its glycogen into glucose through glycogenolysis and releases this glucose to the bloodstream. Also, when muscles exercise, they disintegrate their glycogen through the same process to produce glucose as an energy source. Thus, glycogen serves as a fuel tank that we fill up every time we eat carbs and deplete to keep the motor running.

3-SECOND NUCLEUS
Big carbs of similar structure serve as fundamental energy reserves: starch in plants and glycogen in animals.

3-MINUTE SYSTEM
Although both starch and glycogen are made up of glucose units, they differ in how these are connected: starch contains glucose units in a straight line or with sparse branches, whereas glycogen is branched and bushier than starch. The presence of many branches enables fast glycogenolysis, since this happens by the sequential removal of glucose units from the tips of the branches; the more tips, the faster the breakdown. This facilitates rapid glucose release inside the muscles of a fast-moving animal.

RELATED TOPICS
See also
USING SMALL CARBS
page 78

THE CENTRAL HUB
page 82

ATP FROM AIR
page 86

MOBILIZING THE CELL
page 88

3-SECOND BIOGRAPHIES
CLAUDE BERNARD
1813–78
French physiologist who, in 1857, discovered glycogen in the liver

ERIC HULTMAN &
JONAS BERGSTRÖM
1925–2011 & 1929–2001
Swedish physicians who, in 1966, pioneered the study of how diet and exercise affect glycogen metabolism in human muscle

30-SECOND TEXT
Vassilis Mougios

Big carbs, starch and glycogen provide fuel reserves to drive the body's motors.

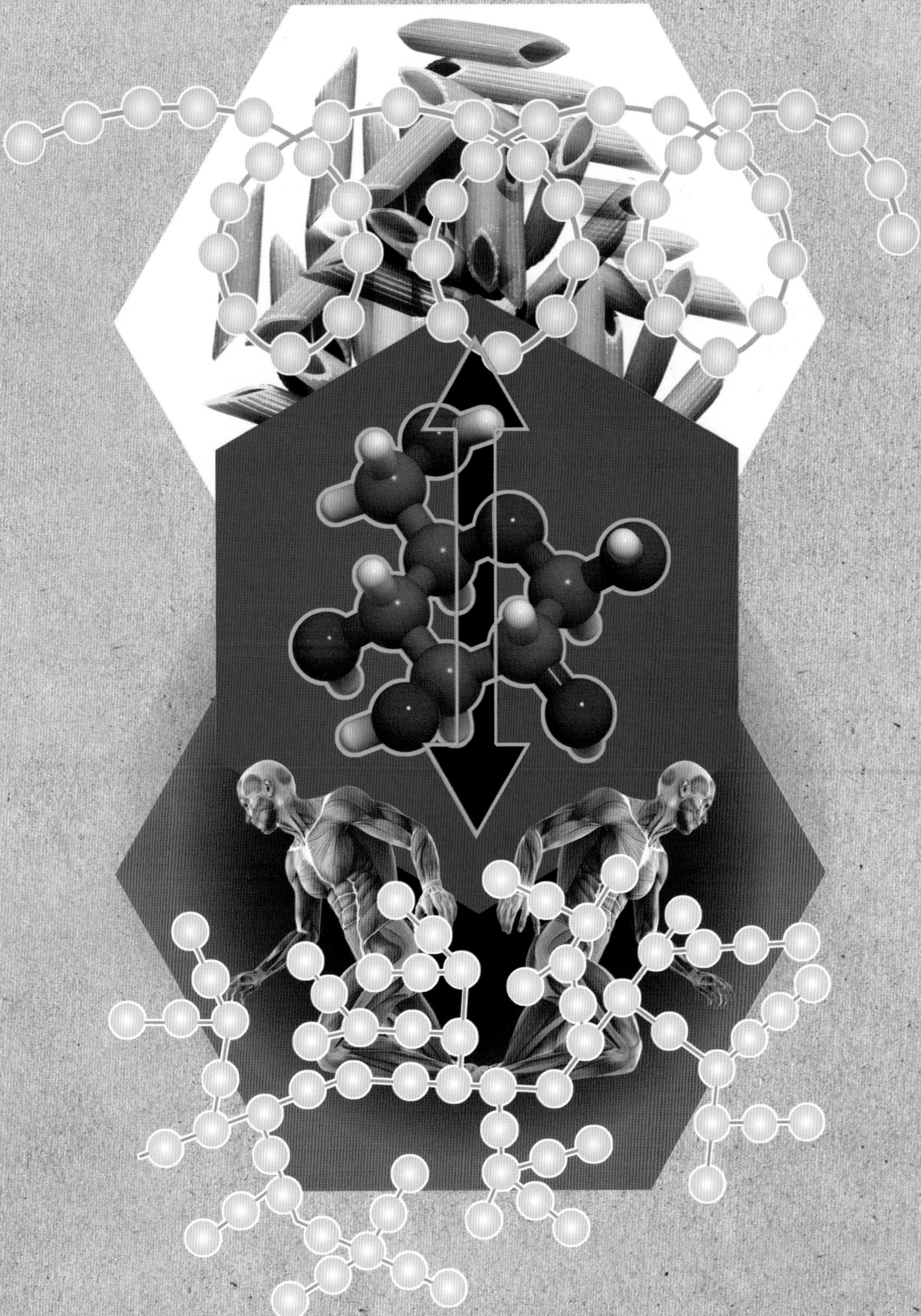

USING
SMALL CARBS

the 30-second structure

Carbohydrates come in many sizes. The smaller ones are monosaccharides and disaccharides. Examples of monosaccharides are glucose, fructose and galactose; examples of disaccharides are sucrose and lactose. Glucose, being the product of photosynthesis in plants, outnumbers all other small carbs in most organisms. In humans, glucose arises primarily from the degradation of starch and glycogen – two polysaccharides (big carbs) found in the diet, liver and muscle. Glucose is a universal energy source: all cells can break it down through glycolysis, a metabolic pathway ending in the compound pyruvate. This makes carbs our prime energy source. Pyruvate stands at the crossroads of two routes, one that uses oxygen (aerobic) and the other that does not (anaerobic). The anaerobic route makes lactate and rapidly produces 2 ATP per glucose (fast energy). The aerobic route burns the glucose to CO_2 and produces 30 ATP per glucose, albeit more slowly (more sustained energy). Apart from breaking down glucose, we can also synthesize it from substances such as amino acids, lactate and glycerol. This is gluconeogenesis; it takes place in the liver and costs 6 ATP per glucose. During hard exercise, the blood transports lactate produced from glucose in muscle to the liver, where lactate is converted back to glucose and refeeds muscle – this is the Cori cycle.

3-SECOND NUCLEUS
Glucose, a small carb, is the top energy source used by all cells; it yields lots of energy fast.

3-MINUTE SYSTEM
Glucose enjoys the privilege of being the exclusive energy source of the brain. When we do not eat enough carbs, the liver compensates for the dearth of dietary glucose by making more, carrying out gluconeogenesis primarily from amino acids at the expense of proteins. This saves the brain – and the rest of the body – from collapse.

RELATED TOPICS
See also
LIFE'S CURRENCY
page 74

USING BIG CARBS
page 76

THE CENTRAL HUB
page 82

ATP FROM AIR
page 86

3-SECOND BIOGRAPHIES
OTTO MEYERHOF
1884–1951
German physician and biochemist who, with other researchers, delineated the glycolysis pathway in the 1930s

CARL CORI & GERTY CORI
1896–1984 & 1896–1957
Czech-American biochemists who, during the 1920s, 1930s and 1940s, described how the body metabolizes glucose, including via their namesake cycle

30-SECOND TEXT
Vassilis Mougios

Small carbs, converted to lactate in muscle, are reborn in the liver and return to muscle.

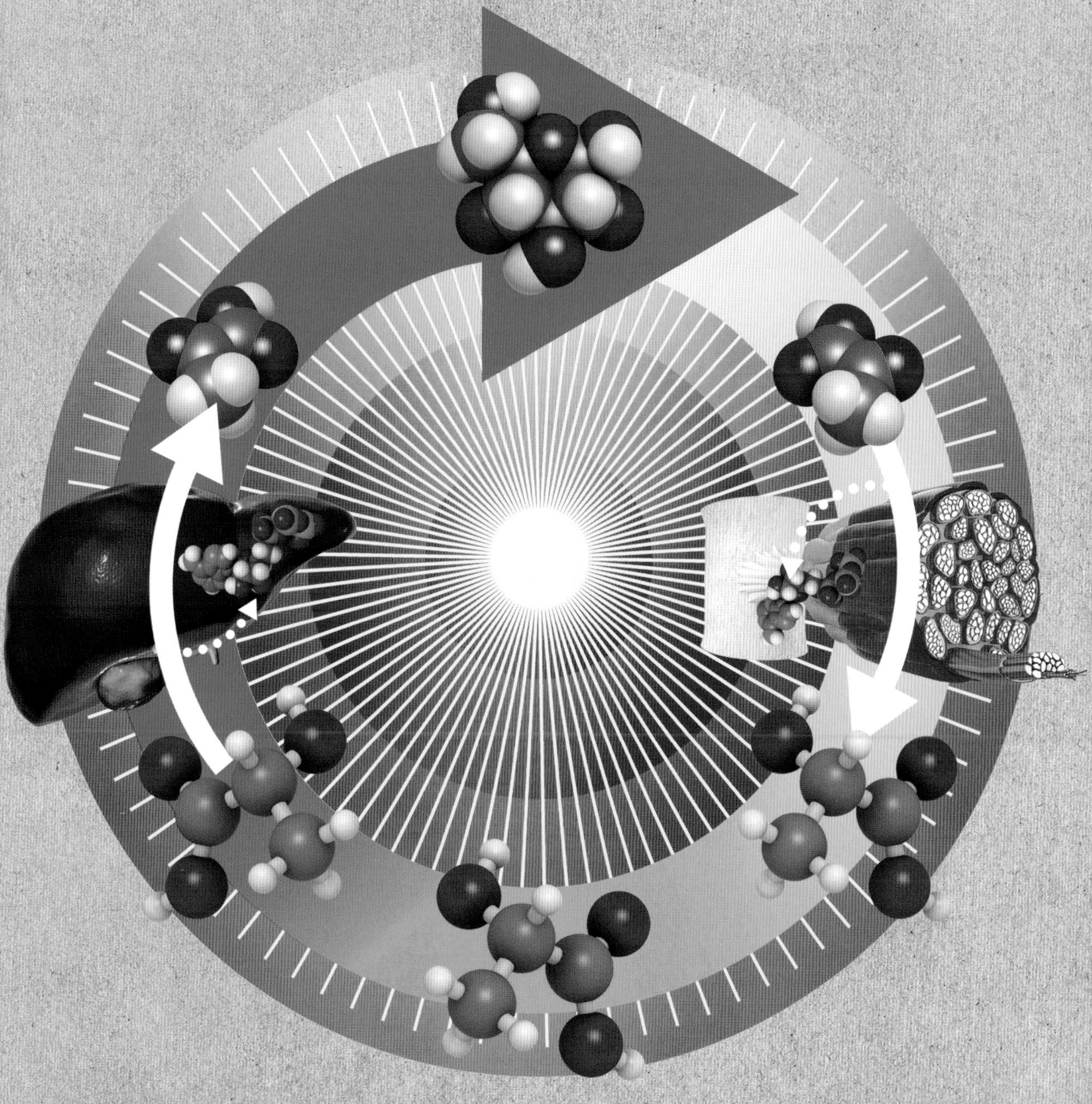

LONG-TERM STORAGE & BURNING FATS

the 30-second structure

RELATED TOPICS
See also
LIPIDS
page 40

LIFE'S CURRENCY
page 74

THE CENTRAL HUB
page 82

Just as in everyday life we do not carry all of our money with us (rather, we deposit it in a bank and use it according to our needs), living organisms do not have all of their energy ready for use. Instead, they mainly draw it from energy storage molecules, of which fat is the most abundant in animals and plants. Fat consists of triglycerides and is found mainly in adipose tissue, a lightweight tissue composed of fat cells called adipocytes. Adipocytes contain big triglyceride droplets; these comprise about 80 per cent of adipose tissue and are the biggest gathering of a substance in a tissue, making triglycerides the most abundant lipid in animals (95 per cent) and food (95–98 per cent). Triglycerides are mainly synthesized from fatty acids and glycerol. This process, known as lipogenesis, leads to a minute increase in adipose tissue. On the other hand, during the increased energy demand of fasting and exercise triglycerides are successively split into fatty acids (lipolysis), broken into two-carbon units called acetyl CoA (beta oxidation) and ultimately burned to yield large amounts of energy as ATP. This causes a minute reduction in adipose tissue. The balance between lipogenesis and lipolysis determines whether we gain or lose fat and, consequently, weight in the long run.

3-SECOND NUCLEUS
Fat has been evolutionarily chosen to be the main energy store in many organisms because it does not weigh much and has high energy yield.

3-MINUTE SYSTEM
The main locations of adipose tissue in the body are under the skin (subcutaneous fat) and around the internal organs (abdominal or visceral fat). Subcutaneous fat is usually more than abdominal fat, with excess of the latter being associated with metabolic disorders. Body fat is the energy reserve with the greatest variation from person to person. People whose body fat comprises more than 30 per cent of their total weight are considered obese.

3-SECOND BIOGRAPHIES
MICHEL EUGÈNE CHEVREUL
1786–1889
French chemist who discovered that fats are triglycerides of fatty acids

CLAUDE BERNARD
1813–78
French physiologist who discovered lipolysis

MARCELLIN BERTHELOT
1827–1907
French chemist who described lipogenesis

30-SECOND TEXT
Anatoli Petridou

Fatty acids are stored in the form of triglycerides – discovered by Michel Eugène Chevreul – until needed for energy.

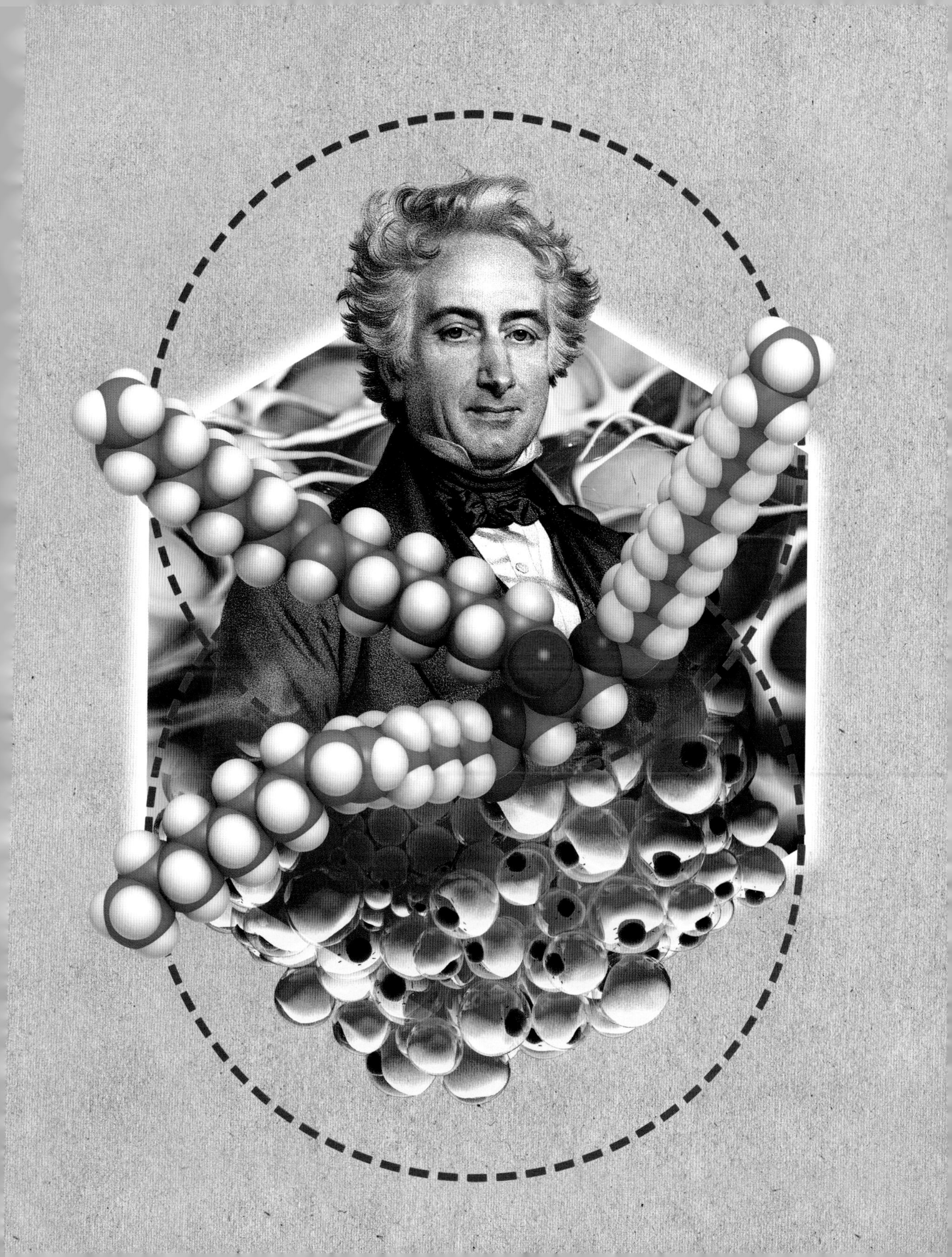

THE CENTRAL HUB

the 30-second structure

The citric acid cycle consists of a series of enzymatic reactions starting and ending at the same biomolecule, citrate (citric acid), after which the cycle is named. It is also known as the tricarboxylic acid cycle (because the first three compounds produced bear three carboxyl groups each) and the Krebs cycle (after Hans Krebs who discovered it). Often referred to as the central metabolic hub, it is the gateway for aerobic energy production from the three major energy sources of cells – carbohydrates, lipids and proteins. The cycle starts with the molecule acetyl CoA, which carries two carbons produced from the oxidation of carbohydrates, lipids and proteins in the form of an acetyl group. When it enters the cycle those two acetyl carbons combine with the four-carbon oxaloacetate to form the six-carbon citrate. In the remainder of the cycle the citrate is successively transformed, eventually into another oxaloacetate, readying the cycle to start again. Along the way, two of the citrate carbons are oxidized into two molecules of carbon dioxide. In this way the cycle produces most of the gas we exhale. In the process, energy-containing compounds are made, which are fed to the electron-transport chain and generate ATP in oxidative phosphorylation.

RELATED TOPICS
See also
USING BIG CARBS
page 76

USING SMALL CARBS
page 78

LONG-TERM STORAGE
& BURNING FATS
page 80

HANS KREBS
page 84

ATP FROM AIR
page 86

3-SECOND BIOGRAPHIES
ALBERT SZENT-GYÖRGYI
1893–1986
Hungarian biochemist who discovered components and reactions of the citric acid cycle

FRITZ ALBERT LIPMANN
1899–1986
German-American biochemist who co-discovered the coenzyme A with Hans Krebs

30-SECOND TEXT
Anatoli Petridou

The Krebs cycle is a fundamental process in aerobic energy production.

3-SECOND NUCLEUS
The citric acid cycle is the central roundabout of metabolism, tirelessly handling the flow of biomolecules and energy from carbohydrates, lipids and proteins.

3-MINUTE SYSTEM
The citric acid cycle is often termed an amphibolic pathway because of its central role in both catabolism (breaking down larger molecules) and anabolism (synthesizing larger molecules). Apart from its catabolic character, intermediates of the cycle serve as substrates for biosynthetic processes. Starting with intermediates of the citric acid cycle, amino acids, glucose and haeme can be produced. So, metabolites of the cycle serve as points of entry or departure for catabolic and anabolic processes.

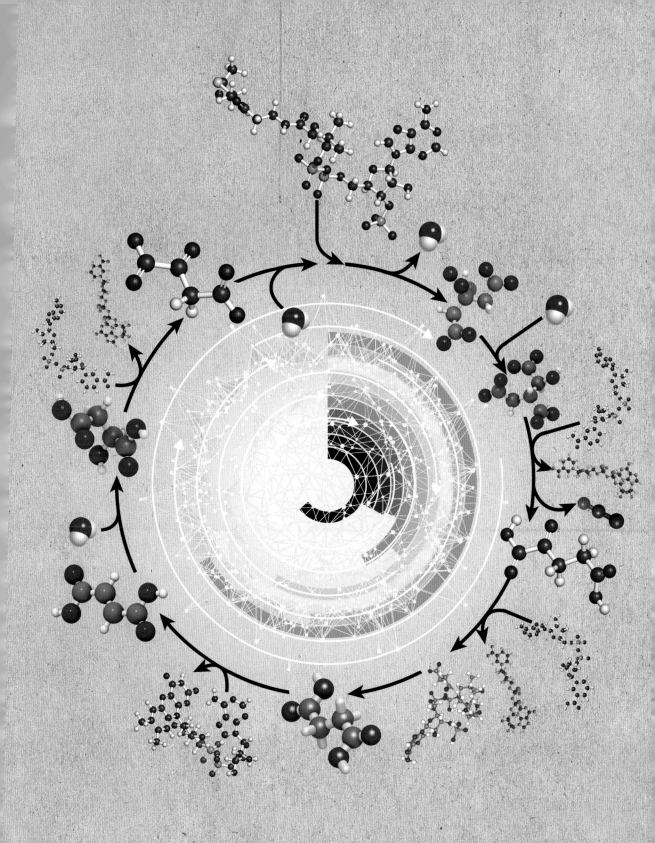

25 August 1900
Born in Hildesheim, Germany

1925
Completes a medical degree from the University of Hamburg

1926
Begins work as an assistant to Otto Warburg at the Kaiser Wilhelm Institute for Biology in Berlin

1932
Discovers the urea cycle that many land animals use to excrete nitrogen

1933
Moves to Cambridge after the Nazi Party rises to power and dismisses persons of Jewish ancestry from academic posts

1937
Elucidates the citric acid or Krebs cycle

1938
Marries Margaret Fieldhouse

1952
Demonstrates the citric acid cycle is a source of biosynthesis precursors

1953
Shares the Nobel Prize in Physiology or Medicine for his discovery of the citric acid cycle

1957
Discovers the glyoxylate cycle, a variant of the citric acid cycle that occurs in some plants and bacteria

22 November 1981
Dies in Oxford, England

HANS KREBS

Even the simplest cell uses a
bewildering variety of metabolic reactions as
it harvests energy, excretes waste and builds
up the molecules of life. The elucidating
of these processes is one of the great
achievements of twentieth-century
biochemistry, made in large part through the
work of Hans Krebs. Krebs discovered several
key pathways, including the citric acid cycle
that lies at the heart of intermediary
metabolism and now bears his name.

Krebs was born in 1900, the middle son of
a physician, in Hildesheim, Germany. After
an austere upbringing, he initially pursued a
career as an ear, nose and throat specialist like
his father. During his medical studies at the
University of Freiburg his interest in metabolism
was awakened when he learned about the
beta-oxidation pathway the cells use to break
down fats from Franz Knoop, who had
discovered the process.

Krebs' early career was a pattern of success
followed by setbacks. After graduating from
medical school, he spent several years
volunteering in research hospitals before
becoming an assistant to Otto Warburg, a
pioneering metabolism researcher who would
later win a Nobel prize for studying how cells
use oxygen. Under Warburg, Krebs learned
how to follow metabolic processes by
measuring the consumption of oxygen and
evolution of carbon dioxide using a special
pressure gauge that Warburg had developed.
Nevertheless, although he was extremely
productive, Krebs had difficulty securing his
first independent position. Eventually, after
being hired by the University of Freiburg, Krebs
discovered the urea cycle, a metabolic cycle that
land animals use to excrete nitrogen. Although
Krebs' postulate of a cyclic pathway was
revolutionary and established his reputation,
shortly thereafter he was dismissed from his
post when the Nazis came to power. Fortunately,
he was able to leave Germany for Cambridge
and eventually secured a position at the
University of Sheffield.

Krebs continued to uncover more of the
reactions involved in nitrogen metabolism while
also expanding his reach to consider other areas
of metabolism. When he discovered that a
product of sugar metabolism called pyruvate
could fuse together with oxaloacetate to give
citrate (via an acetyl CoA, which he didn't know
about at the time), Krebs uncovered the citric
acid cycle of reactions that cells use to oxidize
fuels to carbon dioxide. Sometimes called the
Krebs cycle, it is regarded as the central hub of
metabolism. For its discovery, Krebs won the
1953 Nobel Prize in Physiology or Medicine.

Stephen Contakes

ATP FROM AIR

the 30-second structure

3-SECOND NUCLEUS
Oxidative phosphorylation
is a process whereby
electrons moving through a
series of proteins generate
an imbalance of protons
across a membrane and
produce ATP when the
protons move back.

3-MINUTE SYSTEM
The movement of electrons
within electron transport
chain proteins uses
electrons' wave-like
behaviour. Unlike ordinary
objects, electrons do
not bounce off barriers like
protein chains that they
would not normally be able
to pass through. Instead
they taper off slowly within
the barrier. If the barrier is
low or thin enough the
tapering is incomplete and
the electrons can tunnel
through and come out
the other side. To enable
rapid 'tunnelling', electrons
are typically passed
between sites no more than
a billionth of a metre apart.

Most of the energy animals use
is generated at a thin membrane inside their
mitochondria using high-energy electrons
released from fuel breakdown in the form of the
molecules NADH and $FADH_2$. The electrons are
taken up by membrane-embedded proteins,
which harvest their energy by moving protons
(H^+ ions) across the membrane. To move the
protons, the proteins pass the electrons between
metal-containing sites within the proteins. This
lowers the electrons' energy slightly with each
pass and activates processes that carry protons
across the membrane. When the electrons are
finished with one protein they can be passed to
another with the aid of mobile electron carrier
molecules. The entire process involves a total
of four proteins and two mobile carriers,
collectively called the electron transport chain.
Ultimately the electrons are delivered to
a molecule of oxygen, reducing it to water. In
this way the electron transport chain consumes
the air we breathe. The energy stored in the
form of the proton imbalance is relieved at a
membrane-embedded molecular motor protein,
which uses the flow of protons back across
the membrane to form ATP from ADP and
phosphate ion. This production of ATP from
electrons and oxygen, referred to as oxidative
phosphorylation, generates approximately
1.5 ATP per $FADH_2$ and 2.5 ATP per NADH.

RELATED TOPICS
See also
MOTORS
page 62

LIFE'S CURRENCY
page 74

THE CENTRAL HUB
page 82

BURNING IN REVERSE
page 94

3-SECOND BIOGRAPHIES
PAUL D. BOYER
1918–2018
American biochemist who
proposed the binding change
mechanism by which ATP
synthase works

PETER D. MITCHELL
1920–92
British biochemist who
proposed that proton
imbalance across a membrane
drives mitochondrial ATP
synthesis

30-SECOND TEXT
Stephen Contakes

*Pass the electron – the
electron transport chain
uses oxygen to make
the cell's energy
currency, ATP.*

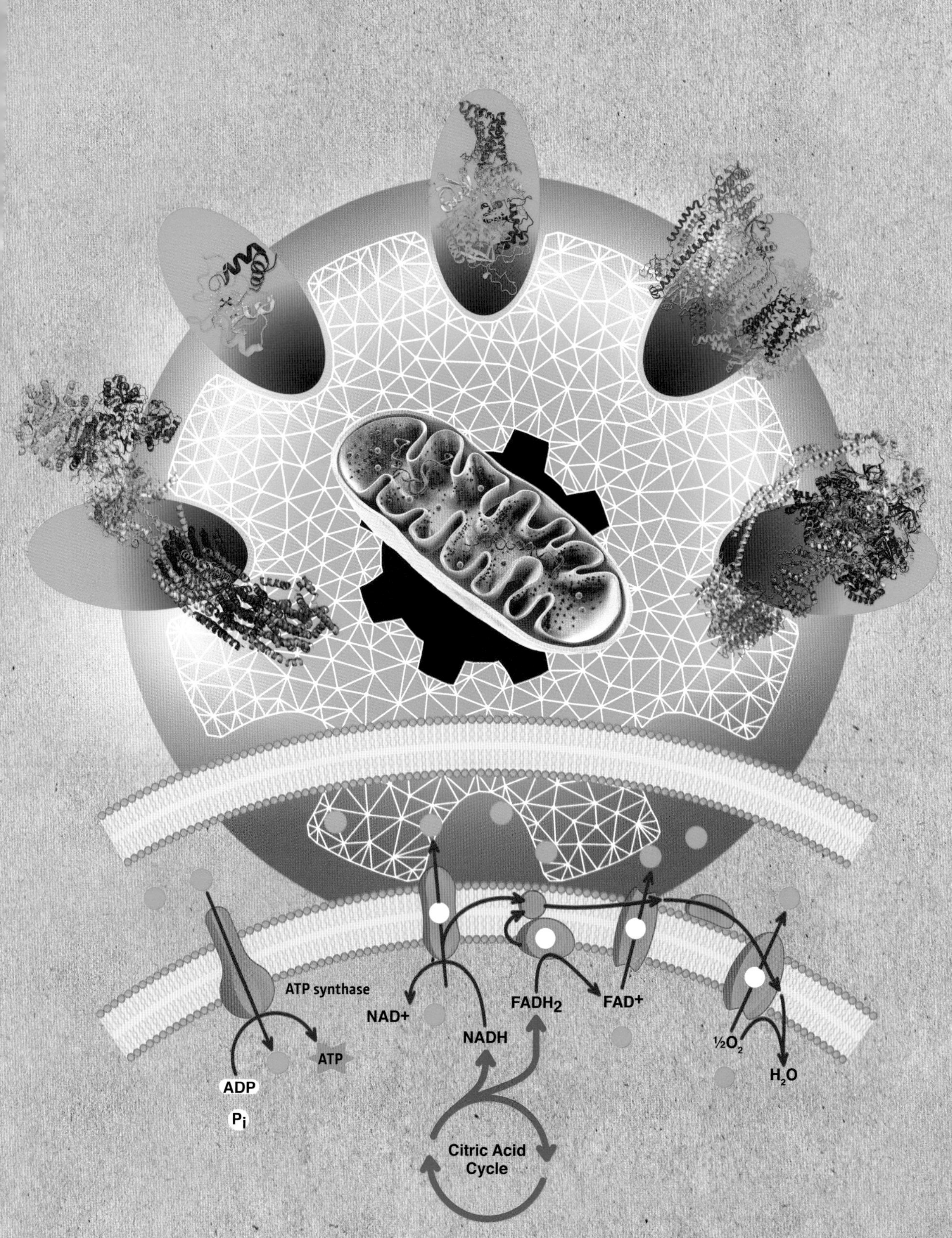
ATP synthase
ADP
Pi
ATP
NAD+
NADH
FADH2
FAD+
½O2
H2O
Citric Acid Cycle

MOBILIZING THE CELL

the 30-second structure

RELATED TOPICS
See also
USING BIG CARBS
page 76

USING SMALL CARBS
page 78

LONG-TERM STORAGE
& BURNING FATS
page 80

After a mixed meal, glucose from carbohydrate digestion, triglycerides from fat digestion and amino acids from protein digestion in the gut flood the bloodstream. The rise in blood glucose triggers insulin secretion from the pancreas. Insulin, a master regulatory hormone, helps glucose enter muscles and adipose tissue, facilitates glycogenesis (glycogen synthesis) in muscle and the liver, promotes lipogenesis (triglyceride synthesis) in adipose tissue and protein synthesis in many tissues. Fasting and exercise trigger the pancreas to release the hormone glucagon. In contrast to insulin, glucagon promotes glycogenolysis (glycogen breakdown) and gluconeogenesis (glucose synthesis) in the liver to release glucose into the bloodstream. Finally, stress, exercise and low blood glucose prompt the nervous system to trigger the adrenal glands to release more adrenaline, which mobilizes the breakdown of muscle glycogen and adipose tissue triglycerides for energy production. All hormones achieve their effects through complex series of molecular interactions termed signal transduction pathways. These amplify the hormones' effects so that one hormone molecule might trigger as many as 10,000 chemical reactions. Transduction pathway malfunction causes diseases such as some types of diabetes, in which inadequate insulin signalling wreaks havoc with metabolism.

3-SECOND NUCLEUS
The nervous and endocrine systems – acting through hormones such as insulin, glucagon and adrenaline – coordinate the efficient storage and use of energy in animals.

3-MINUTE SYSTEM
Most of the hormonal control of energy storage and use rests upon three minute organs – the pancreas and two adrenal glands, which together weigh only 100 g (3½ oz) in humans. Of these, the pancreas secretes both insulin and glucagon. These two hormones have opposite effects, just like the Roman god Janus with his two faces representing peace (insulin, signalling abundance after a meal) and war (glucagon, signalling fasting and exercise).

3-SECOND BIOGRAPHIES
JOHN JACOB ABEL
1857–1938
American biochemist and pharmacologist who, in 1897, isolated adrenaline, the first hormone to be identified

FREDERICK BANTING
& CHARLES BEST
1891–1941 & 1899–1978
Canadian and American medical scientists who, in 1921, discovered insulin and its potential to treat diabetes

30-SECOND TEXT
Vassilis Mougios

Harvesting food's energy requires the concerted action of numerous cells in different organs and tissues.

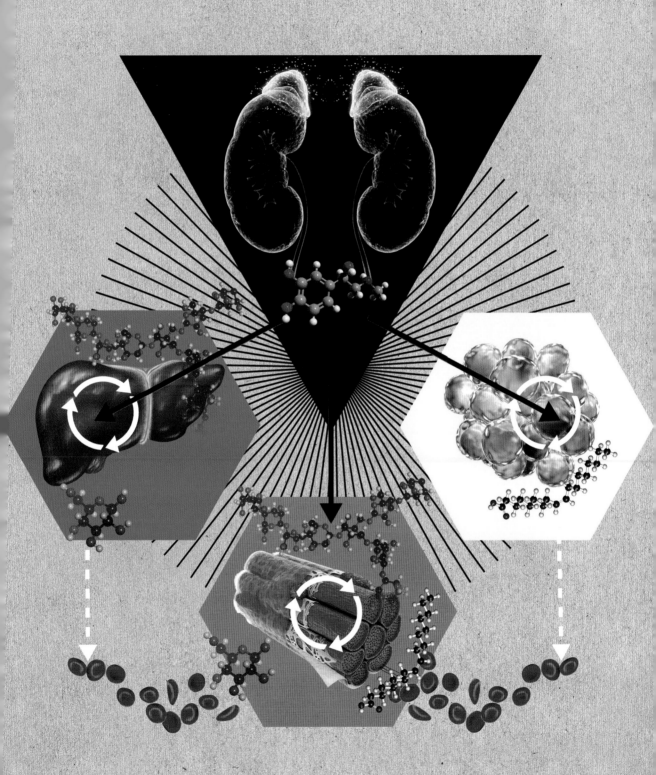

MAKING LIFE ◑

MAKING LIFE
GLOSSARY

action potential The electrical portion of nerve transmission that involves changes in electrical potential down a neuron due to opening and closing ion channels.

bases Also called nucleobases or nitrogenous bases, these are the flat portions of nucleic acids that serve to store and recognize information in biological systems.

chloroplast A green disk-shaped plant organelle in which the light-driven reactions of photosynthesis occur.

chlorophyll A class of green molecules that absorb light in photosystems and use their energy to drive electron flow.

codon A sequence of three DNA or RNA bases that indicates when a protein chain should be started, stopped or extended with a particular amino acid.

denitrification Processes involved in the reduction of nitrate (NO_3^-) that ultimately remove it from biological systems as N_2.

DNA Deoxyribonucleic acid, a type of nucleic acid used to store genetic information.

epigenetic Factors that change how genes are expressed or manifested in an organism because of chemical modifications to DNA or DNA-packaging proteins called histones.

fixation The conversion of an element into a chemical form living systems can use.

genome The complete collection of all of an organism's information-bearing nucleic acids, or genes.

glucogenic Producing glucose. Amino acids are said to be glucogenic if they are degraded in a way that can give rise to glucose.

hydroxylamine A compound formed by adding a hydroxyl or OH to the nitrogen of ammonia, giving a compound of formula NH_3OH.

ion channel A protein that provides a way for ions to move across a biological membrane.

ketogenic Producing ketone bodies – a type of metabolic fuel produced from fats. Amino acids are said to be ketogenic if they are degraded to acetyl-CoA, which can be used to make ketone bodies.

lysosome An organelle consisting of an acidic sac of enzymes that acts as the 'gut of the cell' by catalysing the cleavage of biomolecules.

NADPH Abbreviation for phosphate-bearing nicotinamide adenine dinucleotide hydride, which is used to reduce metabolites in anabolic processes.

neurotransmitter A molecule released into the synapse between neurons that acts as the chemical signal in nerve transmission.

nitrification Processes involved in converting nitrogen fixed as ammonia to nitrate, which plants can more readily absorb.

nucleotide A type of biomolecule consisting of one to three phosphates, a ribose or deoxyribose sugar, and a base. Nucleotides are involved in storing and transmitting genetic information (DNA and RNA), metabolic energy (ATP), electrons (NADH) and chemical signals.

oxidation Chemical processes that remove electrons from atomic centres. In biochemistry this typically involves removing hydrogen atoms or increasing the proportion of oxygen atoms present in a molecule or ion.

phosphate A group of atoms of formula PO_4^{3-} that performs many functions in biomolecules.

proteasome An assembly of proteins that cuts protein chains that have been tagged with ubiquitin into small fragments.

pyruvate A three-carbon metabolic intermediate produced from the breakdown of sugars.

replication The process of copying or reproducing DNA.

RNA Ribonucleic acid, a nucleic acid that serves to transmit the information stored in DNA.

special pair A pair of chlorophyll molecules that receives light energy in photosystems and convert it into electron flow.

transcription The process of converting or 'transcribing' DNA into an RNA message.

translation The process of converting or translating an RNA message into a protein chain.

BURNING IN REVERSE: PHOTOSYNTHESIS

the 30-second structure

3-SECOND NUCLEUS
Plants and photosynthetic bacteria make sugar from CO_2 by using light to make high-energy compounds, which are then used to build CO_2 into sugar.

3-MINUTE SYSTEM
The carbon in your biomolecules comes from photosynthesis. Plants use some of the NADPH and ATP generated using light to incorporate carbon from carbon dioxide into their biomolecules. The most common method involves attaching carbon dioxide to a phosphate-bearing form of the five-carbon sugar ribulose. The protein that catalyses this addition, Rubisco, is the most abundant enzyme on Earth.

Most of life on Earth is powered by sunlight. Yet among living things only plants and a few types of bacteria can harvest light energy directly. They do so using membrane-embedded protein assemblies called photosystems. These contain molecules that absorb light and transfer much of its energy to a special pair of magnesium-containing chlorophyll molecules. An electron within the special pair moves, first within the chlorophyll molecule itself, then through the photosystem and ultimately on to another protein. In the chloroplasts of plants this protein shuttles the electron to a second photosystem, where more light is absorbed to raise the electron's energy even further. After the electron moves through this photosystem it is consumed to make a molecule of NADPH, a phosphate-bearing variant of NADH commonly used in biosynthesis. Just as with the electron transport chain taking place in our mitochondria, electron movement in photosystems serves to carry protons (H^+) across a membrane, generating a proton imbalance that drives the synthesis of ATP. To reset the process, the electrons consumed to make NADH must be replaced. In plants the electrons are taken from two molecules of water, generating four protons and one molecule of oxygen. Thus the photosynthesis of ATP and NADPH also generates the oxygen we breathe.

RELATED TOPICS
See also
LIFE'S POWER
page 18

LIFE'S CURRENCY
page 74

ATP FROM AIR
page 86

3-SECOND BIOGRAPHIES
DANIEL ARNON & ALBERT FRENKEL
1910–94 & 1919–2015
American biochemists who discovered that ATP is produced from light in chloroplasts and photosynthetic bacteria, respectively

MELVIN CALVIN, ANDREW BENSON & JAMES BASSHAM
1911–97, 1917–2015 & 1922–2012
American biochemists who elucidated the Calvin cycle that many plants use to convert CO_2 into sugar

30-SECOND TEXT
Stephen Contakes

Plants use the energy from sunlight to produce glucose from CO_2 and water.

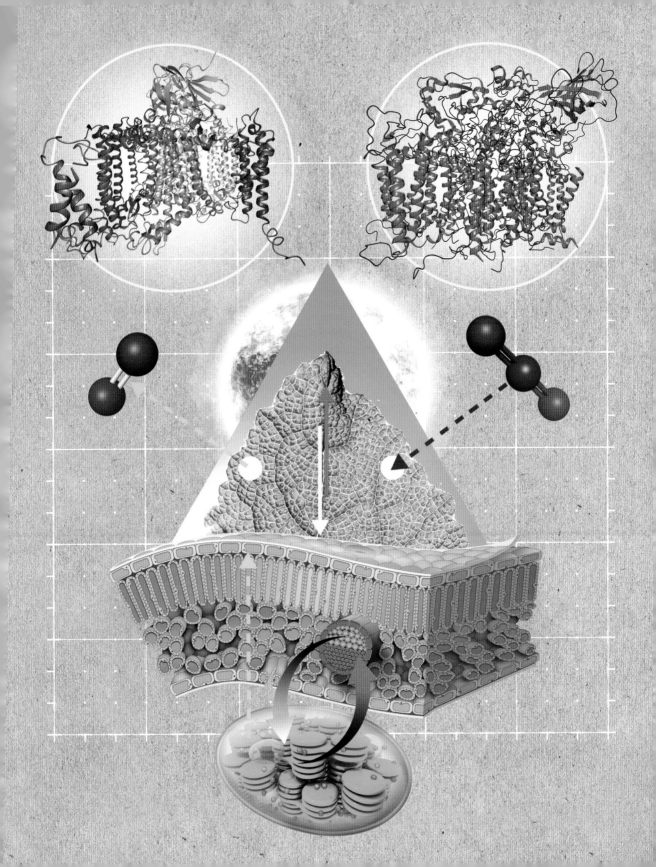

BRINGING IN
THE NITROGEN

the 30-second structure

Nitrogen is an essential element for life and is found in proteins, DNA and some sugars. While nitrogen is abundant – Earth's atmosphere is roughly 80 per cent nitrogen gas (N_2) – the gaseous form is not readily utilized by life. Nitrogen fixation is the process of converting N_2 gas into biologically available nitrogen (primarily ammonia), nitrites and nitrates. The key biologically available nitrogen compound, ammonia (NH_3), is the centre of the nitrogen cycle. While some ammonia is released through decay of organic matter (dead plants, for example) or produced directly ($N_2 \rightarrow NH_3$) by lightning, it is primarily generated from nitrogen gas by bacteria using the enzyme nitrogenase. Plants mainly consume ammonia as a nutrient; however, the predominant mechanism for ammonia consumption – bacterial nitrification – is used to provide some bacteria with energy. Microorganisms called ammonia-oxidizing bacteria consume ammonia mostly via oxygen-dependent processes, first to hydroxylamine (NH_2OH), followed by transformation of hydroxylamine to make nitrite (NO_2^-). Next, other microorganisms called nitrite-oxidizing bacteria convert nitrite to nitrate (NO_3^-). Plants can then absorb nitrate as a nutrient, or bacteria in oxygen-free sediments and soils can convert it to nitrogen gas through a series of chemical reactions called denitrification.

3-SECOND NUCLEUS
The nitrogen paradox: our lungs are filled with it but in a form our bodies cannot use. Microorganisms can provide nitrogen to plants and ultimately animals.

3-MINUTE SYSTEM
For aquarium hobbyists, maintaining nitrogen equilibrium is challenging. Food is introduced to nourish the plants and animals while the plant-to-animal ratio is typically smaller than in nature; thus, nitrates accumulate. Furthermore, poor aeration results in oxygen deficiency and thus build-up of ammonia since its conversion to hydroxylamine is an aerobic process. Ammonia poisoning is the foremost concern: it is believed to permeate fish tissues, where the ionized form (NH_4^+) interferes with biological functions.

RELATED TOPICS
See also
ELEMENTS
page 34

WORKFORCE: AMINO ACIDS & PEPTIDES
page 48

3-SECOND BIOGRAPHIES
MARTINUS BEIJERINCK
1851–1931
Dutch microbiologist who discovered nitrogen fixation in biology

FRITZ HABER & CARL BOSCH
1868–1934 & 1874–1940
German chemists who commercialized ammonia synthesis

30-SECOND TEXT
Andrew K. Udit

Fixed nitrogen in the form of protein is consumed by animals and oxidized to nitrite and nitrate ions by bacteria. Nitrogen balance needs to be carefully maintained in an aquarium.

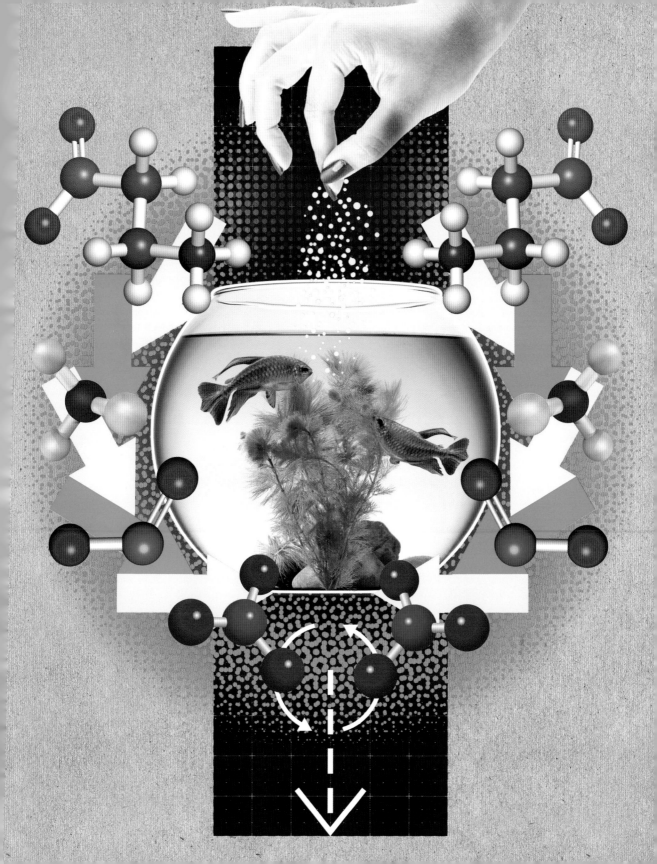

POOLING THE AMINO ACIDS

the 30-second structure

Just as factories keep production flowing smoothly by periodically replacing worn-out machinery, organisms are constantly breaking down and replacing some of their proteins. Within cells this occurs randomly as cellular components become engulfed in organelles called lysosomes and when they are tagged with the regulatory protein ubiquitin and degraded in a multiprotein complex called the proteasome. Both processes split the protein into amino acids, which can be recycled to make new proteins, transformed into useful metabolites such as haems and neurotransmitters or used as a metabolic fuel. To make fuels, the amino acid nitrogen is removed and the remaining 'carbon skeleton' converted to metabolites that can either be built into fats (for amino acids called ketogenic) or carbohydrates (for glucogenic amino acids). In either case the nitrogen removed is excreted, either directly (as in fish) or indirectly in the form of a nitrogen-rich molecule such as urea (as in humans). Many pathways are used to make amino acids, starting with either a sugar metabolism by-product or a citric acid cycle intermediate. Humans cannot perform all of these transformations and need to obtain some amino acids through their diet. Such amino acids are called the essential amino acids, in contrast to the non-essential amino acids we can make.

RELATED TOPICS

See also
WORKFORCE: AMINO ACIDS
& PEPTIDES
page 48

THE CENTRAL HUB
page 82

THINGS WE NEED
page 136

3-SECOND BIOGRAPHIES
DOROTHY HODGKIN
1910–94
British crystallographer who determined the structure of vitamin B12

IRWIN ROSE, AVRAM HERSHKO & AARON CIECHANOVER
1926–2015, 1937– & 1947–
American biologist and Israeli biochemists who discovered how cells selectively degrade proteins using the regulatory protein ubiquitin

30-SECOND TEXT
Stephen Contakes

3-SECOND NUCLEUS
Organisms maintain a pool of amino acids and other nitrogen-containing metabolites by constantly breaking down and synthesizing amino acids in order to meet their metabolic needs.

3-MINUTE SYSTEM
Some diseases are caused by genetic defects that render an amino acid synthesis or breakdown pathway inoperable. Individuals suffering from phenylketonuria cannot make an enzyme needed to break down the amino acid phenylalanine. Vitamin deficiency can have similar effects. Without vitamin B6 the enzymes in your body cannot convert amino acids into the oxygen-binding haem molecule, stopping haemoglobin production and causing you to become anaemic.

Within cells, proteins are broken down in a multiprotein complex called the proteasome.

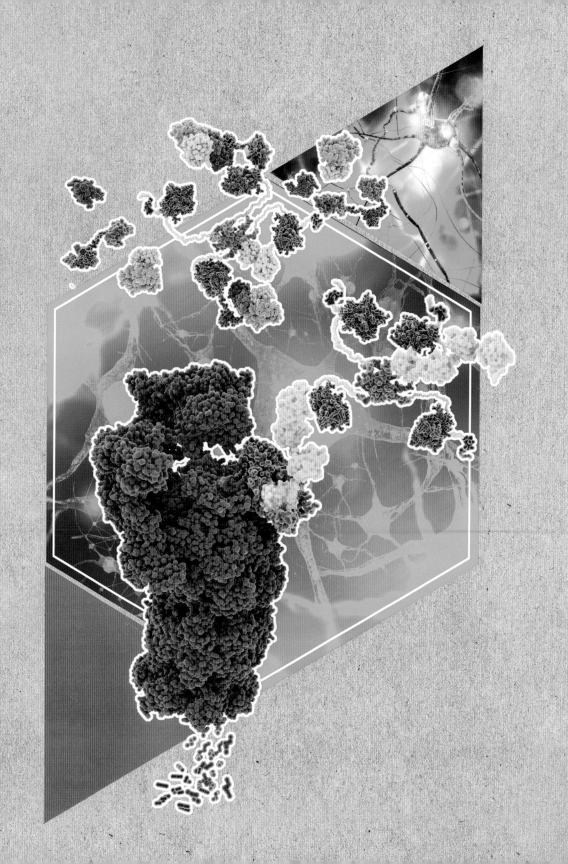

THE MAKING OF THE MESSAGE BEARERS

the 30-second structure

3-SECOND NUCLEUS
The nucleotides that comprise DNA and RNA are made from a phosphate-bearing ribose sugar, amino acids and a few one-carbon units.

3-MINUTE SYSTEM
A modified form of the vitamin folate is used to insert two of the carbons in purines and the carbon that is added to uracil to make the deoxythymidine. When you have a folate deficiency your body cannot make enough nucleotides to replace the cells in your body. This can cause you to become anaemic and increase your risk for colon cancer and cognitive impairment later in life.

Whether as small molecules such as ATP, NADH and $FADH_2$ or as precursors to information-bearing polymers such as DNA and RNA, nucleotides are involved in almost all cellular processes. Cells make them starting from a 5-phosphoribosyl pyrophosphate (PRPP), a ribose sugar bearing multiple phosphates that is made from glucose. Since PRPP already contains the nucleotide's phosphate and sugar, all that remains to make the nucleotide is to attach the flat nucleobases. The way in which this occurs for the bases in DNA and RNA depends on whether they are pyrimidines (Us, Ts and Cs) or purines (As and Gs). Purine bases contain a six-membered hexagon and a five-membered pentagon of atoms fused together at their edges. These are built directly on to the PRPP in 11 steps to give an inosine monophosphate, which is then modified further to make it an A or a G. In contrast, pyrimidine rings contain a hexagon of six atoms and are made as a uridine (U). The preformed uridine is then attached to the ribose to give a uridine monophosphate (UMP), which can then either be used directly or modified into a cytidine monophosphate (CMP). Two additional processes are used to make the deoxyribonucleotides that comprise DNA, the removal of a ribose oxygen (making it 'deoxy') and attachment of a carbon to the uridine, converting the U to a T.

RELATED TOPICS
See also
LIFE'S LETTERS: NUCLEOTIDES
page 46

BRINGING IN THE NITROGEN
page 96

THINGS WE NEED
page 136

3-SECOND BIOGRAPHIES
ALBRECHT KOSSEL
1853–1927
German biochemist who first isolated and identified the nucleobases

JOHN BUCHANAN & DAVID GREENBERG
1917–2007 & 1918–2005
American biochemists who discovered how cells make purine nucleotides

30-SECOND TEXT
Stephen Contakes

The bases of DNA contain single and double rings of atoms called pyrimidines and purines.

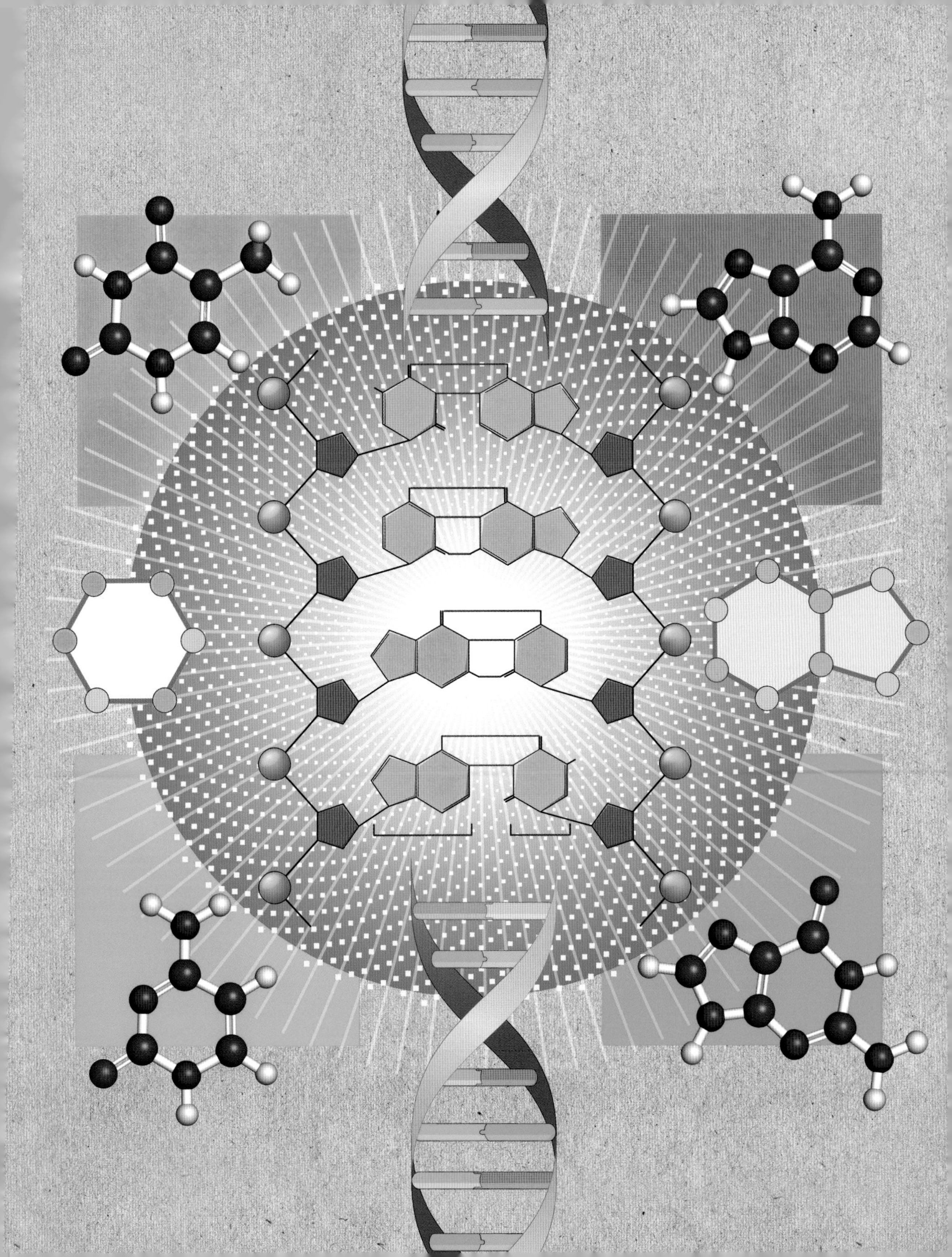

MAINTAINING THE CODE

the 30-second structure

Deoxyribonucleic acid (DNA) assembles into a double helix using a backbone of alternating ribose (a sugar) and phosphate molecules. Between two backbones, the four bases (commonly known as A, T, G and C) are linked by hydrogen bonds: A pairs with T and G pairs with C. The familiar helical twist results from the DNA molecule adopting its lowest-energy conformation. Functionally, the linear sequence of its nucleotide building blocks contains the 'code' that will ultimately be translated into the amino acid sequence of a protein. Each time a cell divides, the entire amount of DNA must be replicated to provide a copy to both daughter cells. The double helix becomes separated between the bases, and each strand serves as a template for the addition of new nucleotides. The replication enzyme DNA polymerase reads each template strand base by base, and links new nucleotides to create a new backbone with bases that pair with those in the template strand. From two original template strands, two identical DNA double helices are created. Part of DNA polymerase activity includes a proofreading function – an incorrectly added nucleotide can be cut out and replaced with the correct one, greatly enhancing the fidelity of DNA replication to approximately one error per 1 million added nucleotides.

RELATED TOPICS
See also
LIFE'S LETTERS:
NUCLEOTIDES
page 46

JAMES WATSON, FRANCES
CRICK & ROSALIND FRANKLIN
page 104

READING THE BOOK
page 106

3-SECOND NUCLEUS
The DNA double helix is designed to faithfully store and copy the genetic code.

3-MINUTE SYSTEM
To appreciate the speed and dexterity of DNA replication using a familiar analogy, consider the replication of the 5 million total base pairs of the common bacterium *E. coli*. Scaled to life size, if DNA polymerase were the size of a delivery truck, packages (added nucleotides) would be delivered at a rate of 1,500 per second while the truck travelled at 600 km/h (375 mph), and the entire 400-km (250-mile) route would be completed in less than 40 minutes – while making only one delivery error!

3-SECOND BIOGRAPHIES
ARTHUR KORNBERG
1918–2007
American biochemist who discovered DNA polymerase and described the basic mechanisms of DNA replication

REIJI OKAZAKI
1930–75
Japanese molecular biologist who discovered that one of the two template strands of a DNA molecule is replicated fragment by fragment rather than continuously, even though both strands are replicated simultaneously

30-SECOND TEXT
Steve Julio

DNA replicates by creating new double helices to form identical 'daughter' strands.

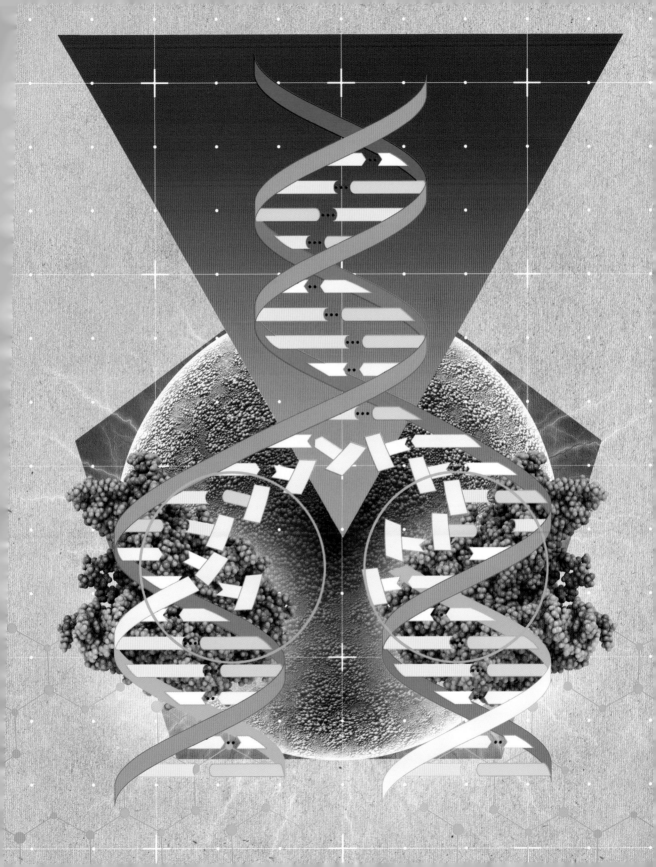

8 June 1916
Francis Crick is born in Northampton, England

25 July 1920
Rosalind Franklin is born in London, England

6 April 1928
James Watson is born in Chicago, Illinois, USA

1952
Franklin (along with research associate Raymond Gosling) produces Photo 51

1953
Watson and Crick publish 'Molecular Structure of Nucleic Acids' in the journal *Nature*

1958
Watson joins the biology faculty at Harvard University

1958
Crick publishes his 'adaptor hypothesis' that correctly described the transfer of information from the DNA code into the amino acid sequence of a protein

16 April 1958
Franklin dies at age 37 in London

1960
Crick becomes research fellow (and later research scientist) at the Salk Institute in San Diego, California

1962
Watson and Crick (and Maurice Wilkins) win the Nobel Prize in Physiology or Medicine; Watson becomes the director of Cold Spring Harbor Laboratory, New York

1968
Watson writes the best-selling book *The Double Helix*, detailing the race for the discovery of the structure of DNA

1990
Watson becomes the first director of the Human Genome Project

28 July 2004
Crick dies at age 88 in San Diego, California, USA

JAMES WATSON, FRANCIS CRICK & ROSALIND FRANKLIN

Molecular biologists James

Watson and Francis Crick (along with Maurice Wilkins) won the 1962 Nobel Prize for determining the structure of DNA. Their success depended largely on the contribution of biophysicist Rosalind Franklin, who produced the clearest indirect 'picture' of DNA using X-ray diffraction. Using data from Franklin's X-ray photo, known as Photo 51, Watson and Crick were able to confirm that the DNA molecule was helical, and had regularly repeating, stacked bases – the precise information required to come up with their famous result that revolutionized the study of biology.

Watson received his PhD from Indiana University in 1950, having spent several years studying the genetics of bacterial viruses before becoming interested in the structure of DNA and joining Crick at Cambridge University. Crick earned his PhD in 1954 from Cambridge, having had his studies interrupted by the Second World War. Crick began his scientific career as a physicist, switching to biology when he became interested in nucleic acids. Watson and Crick enjoyed a fruitful collaboration in part because it combined Watson's creative exuberance with Crick's more methodical disposition.

Franklin earned her PhD in 1945 in physical chemistry from Cambridge University, and by the early 1950s was directing a laboratory studying DNA structure at King's College London, a mere 100 km (62 miles) away from Cambridge University where Watson and Crick were modelling possible DNA structures. The proximity of the two institutions provided the backdrop for a contentious race to discover the structure of DNA.

While Franklin spent numerous hours in the laboratory using X-ray diffraction to produce higher-resolution DNA images, Watson and Crick used available data from colleagues to model potential DNA structures, without conducting experiments themselves. In 1952, Franklin's lab produced Photo 51, which – by means that are still disputed – found its way into the hands of Watson and Crick. Several months later, their famous 1953 paper detailing the structure of the DNA molecule was published in the journal *Nature*.

Watson, Crick and Franklin's work ushered in a new era of biological inquiry focusing on the molecular function of DNA. Their determination of the structure of DNA led directly to the discovery of DNA replication, as well as how the code contained in the sequence of nucleotides is used to direct the synthesis of a protein. Their pioneering work is also the foundation for the current genomics revolution that focuses on sequencing the entire set of DNA in an organism in order to comprehensively study its biology.

Steve Julio

READING THE BOOK

the 30-second structure

The central dogma of the molecular biosciences is this: genetic information flows from DNA to RNA to protein. This occurs as the sequence of bases (As, Ts, Cs and Gs) in a region of DNA is 'transcribed' into a complementary RNA sequence (Us, As, Gs and Cs), which is then 'translated' into the polypeptide chain of a protein. To make the RNA needed, the enzyme RNA polymerase docks with the DNA regions to be copied, exposes the DNA strands by unwinding the DNA double helix and makes RNA by stitching complementary ribonucleotides together. RNA made this way serves several roles. Strands of messenger RNA (mRNA) carry the message to be translated as sequences of three bases called codons. These tell the ribosome when to 'start' or 'stop' making protein and when to add particular amino acids. For example, the codon UGG tells the ribosome to add a tryptophan to the chain. Ribosomal RNA (rRNA) associates with proteins to form large assemblies called ribosomes that catalyse protein formation. They do this using transfer RNAs (tRNAs) that carry amino acids and have 'anticodons' that recognize and stick to the codons of mRNA codons. The tRNAs act as cartridges that fit into ribosomes and line up along the mRNA in the correct sequence, allowing the ribosome to string the attached amino acids together into the polypeptide chain.

RELATED TOPICS

See also
LIFE'S LETTERS:
NUCLEOTIDES
page 46

THE MAKING OF THE
MESSAGE BEARERS
page 100

MAINTAINING THE CODE
page 102

3-SECOND BIOGRAPHIES
MARSHALL W. NIRENBERG
& HAR GOBIND KHORANA
1927–2010 & 1922–2011
American biochemists who
helped establish how DNA
specifies the sequence of
proteins

ROGER D. KORNBERG
1947–
American biochemist who
structurally characterized how
DNA is transcribed to RNA

30-SECOND TEXT
Stephen Contakes

DNA in the nucleus is converted to RNA messages that ribosomes translate into functional proteins.

3-SECOND NUCLEUS
In the conversion of DNA to protein, RNA carries the genetic message, transfers amino acids to the protein chain and helps catalyse the formation of protein chains.

3-MINUTE SYSTEM
In some organisms, including ourselves, the first RNA produced is edited before it is used to make protein. In the same way that video editors remove scenes that don't advance a movie's plot, a RNA-protein complex called the spliceosome makes the mRNA that will be translated into protein by cutting out unwanted RNA regions called 'introns' and stitching the remaining 'exon' regions together. Such splicing also allows several proteins to be made from one RNA strand.

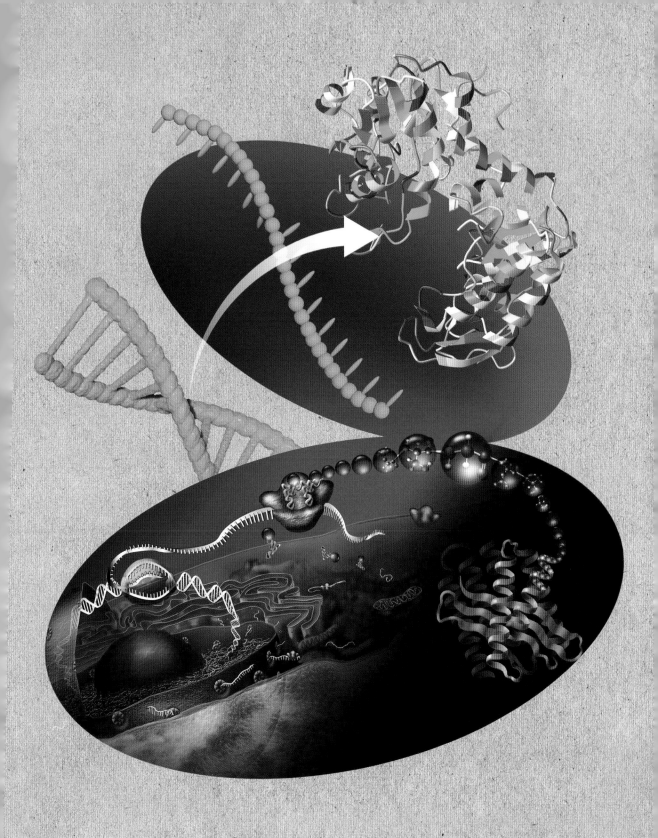

DOCTORING DECODING: EPIGENETICS

the 30-second structure

3-SECOND NUCLEUS
Epigenetic modifications determine how particular cells function by imparting additional information to the main genetic code.

3-MINUTE SYSTEM
An organism's environment can influence epigenetic modification of the DNA, which in turn can influence behaviour. In one experiment, laboratory rat pups that were neglected by their mothers had a different profile of epigenetic modifications on their DNA compared to rat pups who were groomed by their mothers. The neglected rats grew into fearful, withdrawn adults, but littermates who were initially neglected and then allowed to be groomed by a different mother had their epigenetic modifications reversed and grew into normal socialized adults.

Epigenetic modifications are small chemical attachments to the DNA molecule that specify how the main code, contained within the linear sequence of A, T, G and C nucleotides, should be used. These modifications occur at particular places throughout the genome (genetic material) and serve as a chemical tag for DNA-binding proteins to recognize where they should (or should not) bind and then act. For example, an enzyme called DNA methyltransferase can attach a methyl (CH_3) group to cytosine nucleotides in DNA. In some parts of the genome this methyl prevents nearby DNA from being efficiently recognized by the protein machinery that binds to and transcribes it into RNA, precluding it from being translated into protein. Epigenetic modifications are reversible, meaning that the DNA chemical tags can be attached and removed according to cellular needs and responses. The proper placement of all of the modifications at all locations on the DNA coordinates gene expression throughout the genome. Epigenetic regulation is part of the reason why an organism's cells that contain the same DNA can become particular cell types – the epigenomic tags on a muscle cell's DNA are different to those on the DNA of a liver cell, allowing for cell-defining gene expression patterns.

RELATED TOPICS
See also
LIFE'S LETTERS: NUCLEOTIDES
page 46

MAINTAINING THE CODE
page 102

READING THE BOOK
page 106

3-SECOND BIOGRAPHIES
ARTHUR RIGGS
1939–
American geneticist who was one of the first to propose that chemical modification of DNA plays a regulatory role in cellular processes

ADRIAN BIRD
1947–
British geneticist who discovered a key mammalian epigenetic regulatory paradigm based on methylation of C nucleotides and its role in controlling gene expression

30-SECOND TEXT
Steve Julio

DNA is 'tagged' by an epigenetic modification – these biochemical tags determine how the DNA will be used.

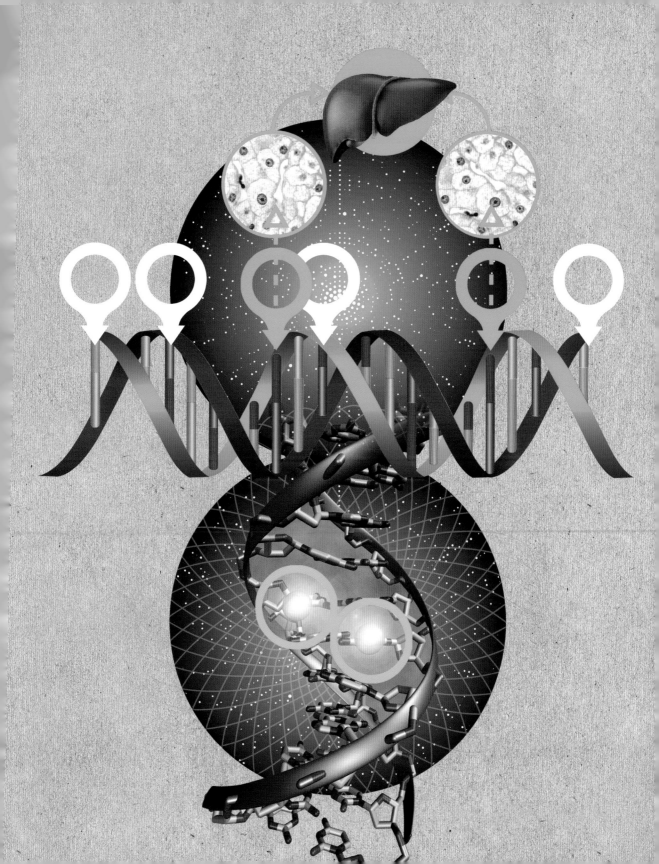

FINDING THE SOURCE

the 30-second structure

3-SECOND NUCLEUS
The origin of life from a non-living planet 4 billion years ago is a mystery, but Earth's chemistry may have shaped primordial biology.

3-MINUTE SYSTEM
Some propose that RNA was the first biomolecule. RNA can both carry information such as DNA and catalyse reactions in the same way as protein. One clue is that the macromolecule that makes protein from RNA, the ribosome, is itself mostly made of RNA, not protein. These hint that an 'RNA world' of replicating RNA-based life may have existed before the development of DNA replication or protein catalysis. Labs are trying to synthesize a self-sustained cycle of RNA reactions to support this hypothesis.

The genetic code shows that all life is related through evolution, but how did this code arise in the first place? Knowing this would relate non-life to life, and geology to biology. The chemical conditions on the early Earth were very different from today, and could have provided the catalysis and energy needed for life. Rock formations acting as chemical reactors could have set up a dynamic cycle of self-replicating and mutating reactions. Some candidates for these reactions have been recreated in laboratories. For example, when the mineral serpentinite is formed, hot seawater is turned into hydrogen, which could fuel life at deep sea vents. At these high pressures and temperatures, sulphur-rich rocks can convert carbon dioxide and that hydrogen into simple molecules such as pyruvate, which are central to life. Asymmetric features on crystal surfaces could have catalysed the synthesis of molecules with a single handedness, as seen in all life. One study found that iron-rich water without oxygen converts glucose and water into many of the intermediate molecules in involved in glycolysis, one of life's fundamental metabolic pathways. We are still far from connecting the dots with a logical series of chemical steps that would start with rocks and end with life, but these experiments hint that such a path might exist.

RELATED TOPICS
See also
LIFE'S POWER
page 18

BONDS: LIFE'S HANDEDNESS
page 36

LIFE'S LETTERS:
NUCLEOTIDES
page 46

READING THE BOOK
page 106

3-SECOND BIOGRAPHIES
HAROLD UREY & STANLEY MILLER
1893–1981 & 1930–2007
American chemists who ran the first experiments showing that simple chemicals under early Earth conditions can synthesize chemicals important for life

ALEXANDER RICH
1924–2015
American biochemist who first proposed what became known as the 'RNA World' hypothesis

30-SECOND TEXT
Ben McFarland

Did chemical reactions in rock formations on early Earth produce the building blocks of life?

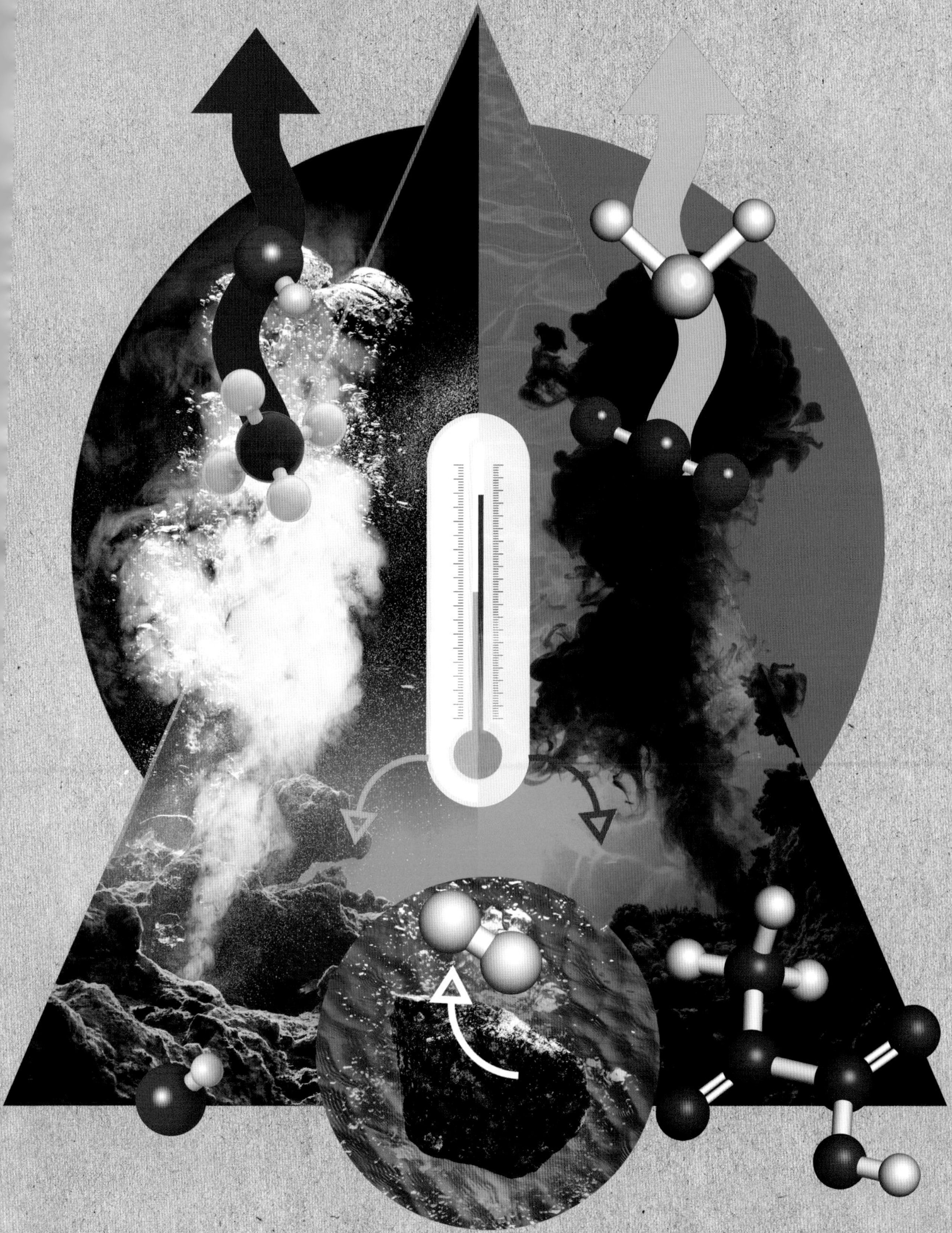

STAYING ALIVE

STAYING ALIVE
GLOSSARY

actin A protein that aggregates together with other actin proteins to form the thin filaments in muscle.

action potential A change in the electrical potential across the membrane of a neuron as it transmits a signal across its surface.

ammonia A molecule of formula NH_3, the protonated form of which, ammonium ion or NH_4^+, is a common but toxic form in which nitrogen is found in biological systems.

antigen A molecule or part of a molecule to which an antibody binds.

carbonic acid A molecule formed from carbon dioxide and water of formula H_2CO_3, which acts as an acid.

depolarization The loss of a charge imbalance – and its associated voltage or polarity – across a cell membrane, such as that which occurs when an electrical signal travels through a neuron as an action potential.

dopamine A neurotransmitter involved in many physiological functions, notably movement, pleasure, mood and cognition.

hydrocarbon A molecule or part of a molecule comprised only of hydrogen and carbon; it is hydrophobic and can be oxidized to release lots of energy.

ion channel A protein that provides a way for ions to move across a biological membrane.

myofibril A cylindrical array of stacked actin and myosin filaments inside a muscle fibre.

myosin A protein that aggregates together with other myosin proteins to form the thick filaments in muscle. It contains a 'head' that binds to actin, where it uses ATP-driven 'swinging' to drive muscle contraction.

neuron Another term for a nerve cell, which consists of a central body surrounded by branching dendrites and a long axon tail. A neuron transmits signals electrically as action potentials moving across its cell membrane and chemically as neurotransmitters are released into the synapse between the end of its axon and neighbouring cells.

neurotransmitter A molecule released into the synapse between neurons that acts as the chemical signal in nerve transmission.

nephrons The basic functional units of the kidneys that serve to filter waste out of the blood and process it for excretion as urine.

oxidation Chemical processes that remove electrons from atomic centres. In biochemistry this typically involves removing hydrogen atoms or increasing the proportion of oxygen atoms present in a molecule or ion.

pathogen A disease-causing organism or virus.

pH A measure of the acidity of a solution. Lower pH values correspond to more acidic solutions.

polysaccharides Carbohydrate polymers that are made by linking many monosaccharide units together.

serotonin A neurotransmitter involved in regulating many physiological functions including cognition, learning, memory and mood. Since serotonin levels are associated with feelings of well-being it is sometimes called the happiness neurotransmitter.

synapse A tiny gap in between the axon of one neuron and the dendrites of the next into which neurotransmitters are released when a signal is transmitted from one neuron to another.

triglycerides Lipid major component of animal fats and plant oils that consists of three fatty acids linked to a single glycerol unit.

villi Small projections extending from the surface of biological membranes, particularly in the intestines, which give the membrane the appearance of a shaggy rug and serve to increase the area available to hold transporters that can move molecules or ions across the membrane.

IN & OUT: DIGESTION & EXCRETION

the 30-second structure

Your body takes up most of the food you digest in the intestines using cells that can absorb salts and small water-soluble molecules such as glucose. In contrast, larger molecules such as proteins, polysaccharides, nucleic acids, and triglycerides must be broken down into their components. This is done by a variety of digestive enzymes, starting in the mouth as amylase begins trimming bush-like starch molecules into smaller saccharides. Protein digestion begins in the stomach as its acid causes the proteins to unfold. This exposes the bonds holding them together to gastric peptidase enzymes, which cut them into small peptide chains. Further food breakdown occurs in the intestine, where enzymes produced in the intestine and pancreas help produce molecules your intestinal villi can absorb. In the intestines, peptide and digestible saccharide chains are broken into smaller fragments and ultimately individual amino acids and monosaccharides such as glucose. Nucleic acids such as DNA and RNA are broken down first into nucleotides and later into phosphate, carbohydrates and nucleobases. The large triglyceride-rich fat portions of your food are dispersed into small droplets coated with the greasy ends of bile acids. Then fatty acids are cleaved off the triglycerides until only one remains. After this any remaining large biomolecules are excreted.

3-SECOND NUCLEUS
The carbohydrates, proteins, nucleic acids and lipids in food are broken into their components by enzymes before being taken up by cells.

3-MINUTE SYSTEM
The bacteria in your digestive track help you digest your food and stay healthy. They metabolize carbs and fats that we cannot; make useful molecules such as folic acid, vitamin K, arginine and glutamine; defend us from dangerous bacteria; and, adjust the composition of the bile produced by liver. For these reasons intestinal bacteria are sometimes considered the 'forgotten organ'.

RELATED TOPICS
See also
LIPIDS
page 40

SWEETNESS: CARBOHYDRATES
page 42

WORKFORCE: AMINO ACIDS & PEPTIDES
page 48

DETOX
page 126

3-SECOND BIOGRAPHIES
GOTTLIEB SIGISMUND CONSTANTIN KIRCHHOFF
1764–1833
Russian chemist who first discovered enzymatic starch digestion

JOHN HOWARD NORTHROP
1891–1987
American biochemist who proved that digestive enzymes were proteins

30-SECOND TEXT
Stephen Contakes

Food is broken down into increasingly smaller units before being absorbed or excreted.

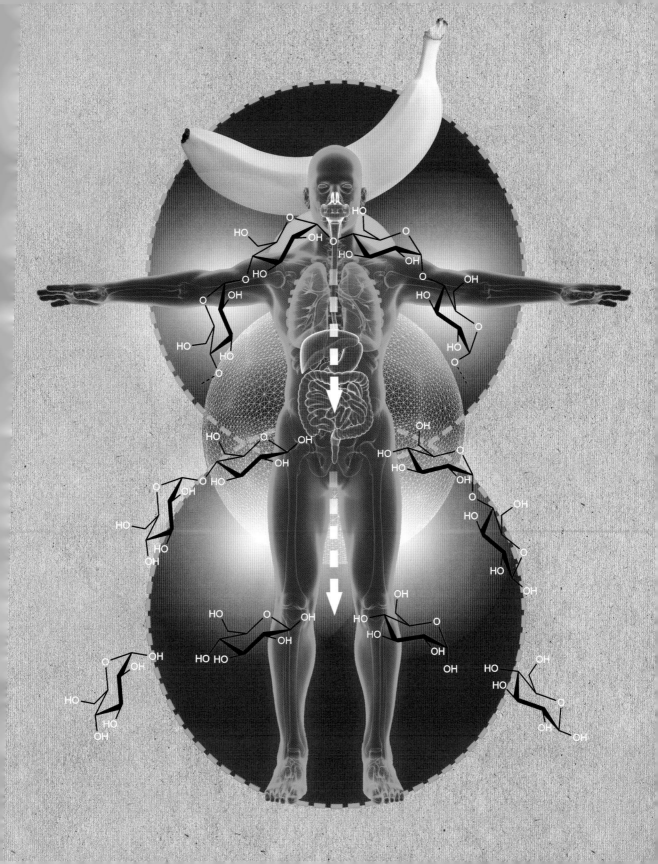

BLOOD, BREATH & BALANCE

the 30-second structure

3-SECOND NUCLEUS
The blood moves the chemicals of life to where they can be used and keeps the acidity, temperature and distribution of water throughout the body in constant balance.

3-MINUTE SYSTEM
The balance of blood molecules is maintained by energy-intensive processes. These stop at the time of death, and then molecules leak out of the lifeless cells. By measuring the concentrations of molecules such as urea and iron in the blood, a coroner might be able to determine the time of death down to the hour.

Our bodies must keep the levels of many chemicals in balance as we eat food and excrete waste. The bloodstream helps maintain this balance as it moves ions and molecules around the body. The water molecules of blood dissolve salts and some molecules, enabling the blood to move them easily. Some crucial molecules do not dissolve but instead are transported by soluble transport molecules that do. For instance, haemoglobin present in red blood cells binds oxygen in the lungs and carries it through the bloodstream to the tissues. When the oxygen is used to break down fuels, haemoglobin carries the carbon dioxide produced from tissues to the lungs. In facilitating the removal of carbon dioxide from our tissues, the blood helps maintain healthy cellular acidity by preventing the formation of harmful levels of carbonic acid. Excess hydrogen ions from the acidic, oxidized metabolites produced as the body oxidizes sugar for energy are further kept in check when they are absorbed by the carbonate present in the blood, which along with carbonic acid acts to maintain blood's pH. Finally, so that nitrogen does not dissolve in the blood as toxic ammonia ions, the blood transports nitrogen produced in the tissues safely as the amino acid glutamine and converts it to the less dangerous molecule urea for excretion.

RELATED TOPICS
See also
LIFE'S MATRIX
page 22

LIFE'S GIVE & TAKE
page 24

TRANSPORT & STORAGE
page 64

ELECTROLYTE BALANCE
page 128

3-SECOND BIOGRAPHIES
JOHN SCOTT HALDANE
1860–1936
Scottish physiologist who discovered that oxygen binding to haemoglobin causes carbon dioxide to dissociate, called the Haldane effect

JEROME KASSIRER
1932–
American doctor who, with Howard L. Bleich, developed a modified form of the Henderson–Hasselbalch Equation, useful for calculating carbon dioxide concentrations in the lungs

30-SECOND TEXT
Ben McFarland

Blood is vital for the transport of oxygen from the lungs.

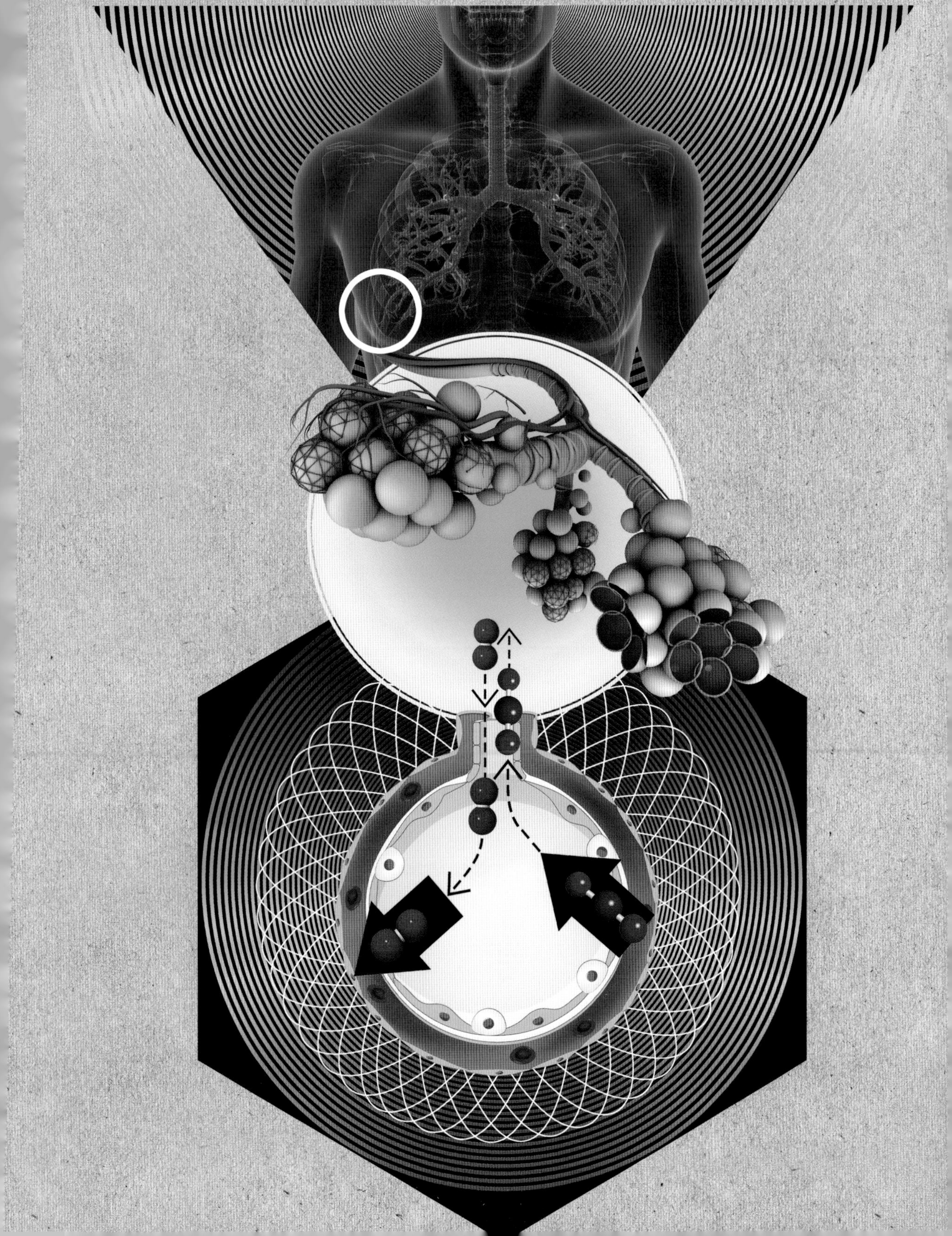

MOVING

the 30-second structure

3-SECOND NUCLEUS
Muscles endow animals
with movement thanks
to complex interactions
between proteins such
as myosin, actin,
tropomyosin and troponin
– interactions controlled
by the nervous system.

3-MINUTE SYSTEM
Muscle relaxation is as
important as muscle
contraction. When the
nervous system stops firing
signals, acetylcholine stops
causing the release of Ca^{2+}
around the thin filaments.
Calcium is taken back
up into its reservoir, a
membranous system called
sarcoplasmic reticulum.
The lack of Ca^{2+} causes
troponin to revert to a form
that holds tropomyosin
in a position hindering
actin–myosin interaction.
When myosin can no
longer bind to and pull
actin, relaxation ensues.

Movement is an essential part of
all forms of life, but it is nowhere more evident
than in animals. Animals owe movement to three
kinds of muscle: skeletal, cardiac and smooth.
Of those, skeletal muscles are the ones moving
us around and are the most highly organized
and versatile. The cells of skeletal muscle, or
muscle fibres, contain bundles of long threads
called myofibrils resembling spaghetti in a
pack. Each myofibril, in turn, consists of thick
filaments, containing the protein myosin, and
thin filaments, composed of the proteins actin,
tropomyosin and troponin. Two other proteins,
nebulin and titin, stabilize the myofibril
structure. Movement starts with the active
sliding of the thin filaments past the thick ones,
resulting in the shortening of myofibrils, muscle
fibres and whole muscles. This sliding is due to
'heads' in the myosin structure, which form
cross bridges to the actin and pull it when ATP is
split on the myosin heads, converting chemical
into mechanical energy. Thus, myosin is a
molecular motor running on ATP. The nervous
system controls muscle activity by releasing
acetylcholine (a neurotransmitter) from motor
nerve endings. Acetylcholine triggers the release
of calcium ions (Ca^{2+}) around the thin filaments,
where Ca^{2+} binds to troponin. This causes
tropomyosin to move away from the myosin-
binding sites on actin and permits contraction.

RELATED TOPICS
See also
LIFE'S CURRENCY
page 74

BIOCHEMISTRY ON THE BRAIN
page 124

STAYING FIT
page 138

3-SECOND BIOGRAPHIES
VLADIMIR ENGELHARDT
& MILITSA LYUBIMOVA
1894–1984 & 1898–1975
Soviet biochemists (husband
and wife) who, in 1939,
discovered that myosin splits
ATP to produce muscle energy

SETSURO EBASHI
1922–2006
Japanese physiologist who, in
the 1960s, discovered troponin
and the control of muscle
activity by Ca^{2+}

30-SECOND TEXT
Vassilis Mougios

*Muscles contract
thanks to the sliding
action of thick myosin
filaments and thin actin
filaments triggered by
ATP breakdown.*

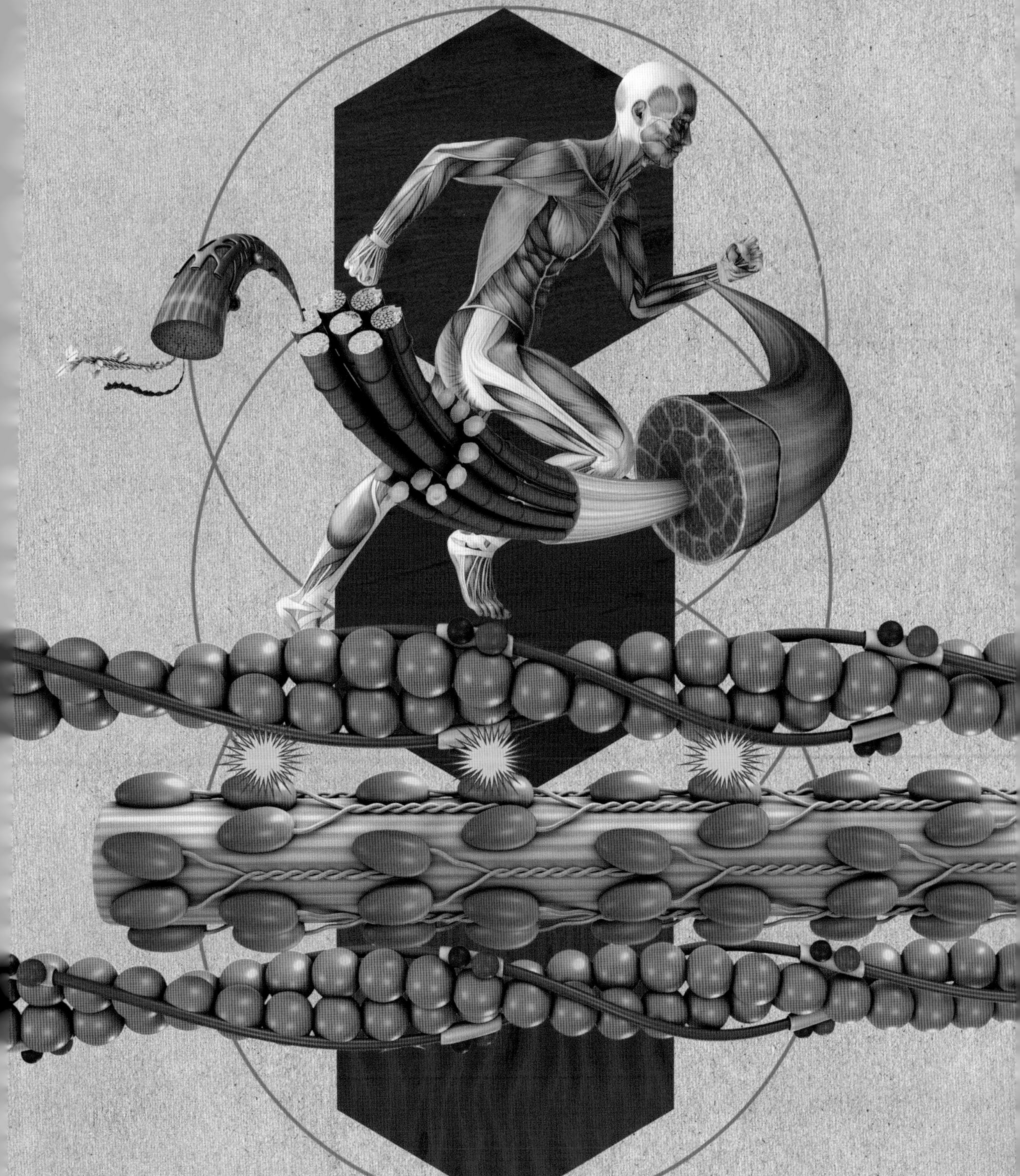

25 February 1924
Born in Birkenhead,
Cheshire, England

1941
Huxley begins
undergraduate work in
physics at Christ's
College, Cambridge

1943
Huxley develops
improved height-to-
surface (H2S) radar
systems during his
wartime service in the
Royal Air Force, an
achievement recognized
by the award of an Order
of the British Empire
(OBE)

1949
Huxley begins studying
muscle during PhD work
in biophysics under John
Kendrew at Cambridge

1952
Huxley begins studying
muscle contraction using
electron microscopy while
completing postdoctoral
work at Massachusetts
Institute of Technology
(MIT) with Frank Schmitt

1954
With Jean Hanson,
Huxley publishes the
sliding filament model
of muscle structure.
Concurrently, Andrew
Huxley (no relation) and
Rolf Niedergerke propose
sliding filaments based
on very different
evidence

1958
First proposes the
'swinging cross-bridge
hypothesis' according to
which ATP powers muscle
contraction by driving
the swinging of bridges
between the actin
and myosin

1966
Marries Frances Maxon
Fripp

1981
Demonstrates the
swinging of cross bridges
by measuring how the
intensity of scattered
X-rays changes as the
bridges swing

25 July 2013
Dies in Woods Hole,
Massachusetts, USA

HUGH HUXLEY

One aim of biochemistry is to relate the behaviour of things such as cells, organs and organisms to the properties of atoms and molecules. This is a difficult task and one not easily accomplished. This is well illustrated by muscle contraction, which had already been studied for more than a century by the time Hugh Huxley began his nearly 60-year quest to explain how muscles work in the 1940s.

Huxley was born in 1924 in Birkenhead, Cheshire, England. After an interlude of military service during the Second World War, in 1947 he completed undergraduate work in physics at Cambridge University. Left unenthusiastic about atomic physics in the wake of the atomic bombing of Japan, he switched to biophysics for his graduate work and began studying the structure of muscle.

When Huxley started, nobody knew exactly how muscle worked. It was known that the appearance of pieces of muscles called sarcomeres change when muscle contracts and that muscles contain filaments of the proteins actin and myosin. By passing X-rays through thin slices of muscle, Huxley was able to determine that the actin and myosin fibres were interlaced: each actin fibre was surrounded by a hexagonal array of six myosin and each myosin fibre by a similar array of six actin. From differences in the X-ray scattering pattern produced by resting and contracting muscle, Huxley hypothesized that bridges form between actin and myosin during muscle contraction.

Realizing he would need additional structural evidence to establish the existence of cross bridges, after completing his PhD Huxley joined the lab of Frank Schmitt at MIT. Schmitt was an expert in electron microscopy, which can 'see' muscle in much finer detail than light microscopy. There Huxley developed a way to make muscle cross sections thin enough to visualize well and began collaborating with Jean Hanson, an expert in light microscopy who had earlier demonstrated how muscle's structure changes on exposure to ATP. Using dilute ATP solutions, the two measured the formation and movement of dark 'bands' in muscle's structure during contraction. Attributing the bands to the thicker myosin filaments, they proposed that the actin and myosin filaments slide past one another during ATP-driven contraction. Nevertheless, several decades passed between their proposal of this 'sliding filament theory' in 1954 and its full acceptance, during which Huxley and others amassed an increasingly compelling array of evidence for both the sliding filaments and Huxley's hypothesis that actin and myosin 'row' past one another using ATP-powered 'swinging cross bridges'.

Stephen Contakes

BIOCHEMISTRY ON THE BRAIN

the 30-second structure

3-SECOND NUCLEUS
Signals move through the brain's nerves in electrical and chemical waves, as ions flow inside each cell and neurotransmitters move from cell to cell.

3-MINUTE SYSTEM
The sodium-potassium ATPase ion pump that creates the transmembrane voltage might be the most important protein in the body. Copies of this protein consume approximately one-quarter of the ATP animals use, similar to how a furnace consumes fuel to keep a house warm. If environmental changes stress the cells, this hard-working protein has to work even harder. Sea urchins grown in higher levels of carbon dioxide, simulating climate change, change their energy budget to consume about 45 per cent of their ATP with this one protein.

The process of thinking a thought sends signals from one part of the brain to another. These signals are both electrical and chemical in nature, and are mediated by flowing ions with negative or positive electrical charge. Resting neurons (nerve cells) have more negative ions on the inside and positive ions on the outside because of the action of an ion pump called the sodium-potassium ATPase that moves positive charge outside. This electrical imbalance can be measured as a voltage across the neuron membrane. When neurotransmitters (such as dopamine or serotonin) are detected at one end of a neuron, they trigger sodium-ion permeable channels to open in the cell membrane. The open channels allow positive sodium ions to flow into the neuron, neutralize the negative charge inside and eventually make the inside positive. This process, called the action potential, triggers other ion channels to open in order, down the length of the cell, until the wave of depolarization reaches the other end. There, the flowing positive ions start a chain reaction that releases stored chemicals, called neurotransmitters, from the end of the neuron into the synapse, where they trigger ion flows that can be detected by the next neuron and the action potential can recur.

RELATED TOPICS
See also
MEMBRANES, MESSENGERS & RECEPTORS
page 66

LIFE'S CURRENCY
page 74

3-SECOND BIOGRAPHIES
SIR JOHN ECCLES
1903–97
Australian physiologist who measured neural excitation as an electrical voltage

ALAN HODGKIN & ANDREW HUXLEY
1914–98 & 1917–2012
British biophysicists who found the mechanism of voltage-dependent sodium channels

JENS SKOU
1918–2018
Danish biochemist who discovered the sodium-potassium ATPase

30-SECOND TEXT
Ben McFarland

A single thought is part of a complex electrochemical signalling process inside the brain's nerve cells.

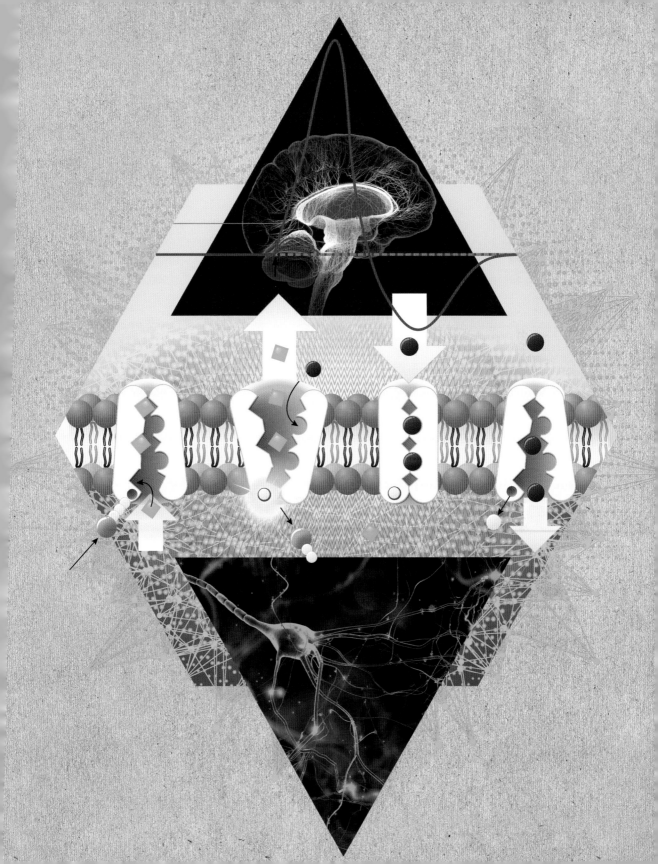

DETOX

the 30-second structure

Supposedly, 'you are what you eat' – and drink – however, your liver may have something to say about that. Indeed, that is a good thing: not everything we eat and drink is good for us, and one critical role your liver plays is in detoxification. Central to this role are enzymes called cytochrome P450s. Simply known as P450s, these enzymes are found throughout the body and primarily in the liver. P450s perform a variety of chemical reactions, the most famous of which is oxidation. Critical for the function of P450 is the iron ion at its reaction centre, which it uses to transfer electrons and oxygen to the molecule it is acting on. P450 oxidation most often results in the insertion of an oxygen atom into a molecule. This typically converts a hydrocarbon part of a molecule to an alcohol group, making the molecule less greasy and able to more easily dissolve in water. This facilitates the elimination of toxins in urine. An example: the char on your over-cooked steak will contain hydrocarbons, such as the carcinogen pyrene, which do not easily dissolve in water. Once in the liver, P450s oxidize these hydrocarbons so they more readily dissolve in urine to expedite their excretion.

RELATED TOPICS
See also
LIFE'S MATRIX
page 22,

MAKING LIFE GO: ENZYMES
page 68

IN & OUT: DIGESTION
& EXCRETION
page 116

3-SECOND BIOGRAPHIES
RICHARD WILLIAMS
1909–79
Welsh biochemist who first described metabolism of xenobiotics by (then unknown) P450s

TSUNEO OMURA
1930–
Japanese biochemist who first identified the enzyme as cytochrome P450

30-SECOND TEXT
Andrew K. Udit

3-SECOND NUCLEUS
From making hormones to activating drugs, P450 enzymes can do it all, including acting as the liver's refuse collectors to help eliminate toxins.

3-MINUTE SYSTEM
Sometimes P450s are too efficient. The drug paracetamol is also oxidized by P450, which further reacts to make toxins. In small doses these toxins are dealt with by the liver (scavenger molecules such as glutathione round them up); however, excessive amounts can lead to liver damage. And pay attention to those warning labels about not consuming alcohol with paracetamol – alcohol induces your liver to make more P450s, which increases the chances of paracetamol toxicity.

The liver's refuse collectors, P450 enzymes play a crucial role in detoxification – as well as the oxidation of medications such as paracetamol.

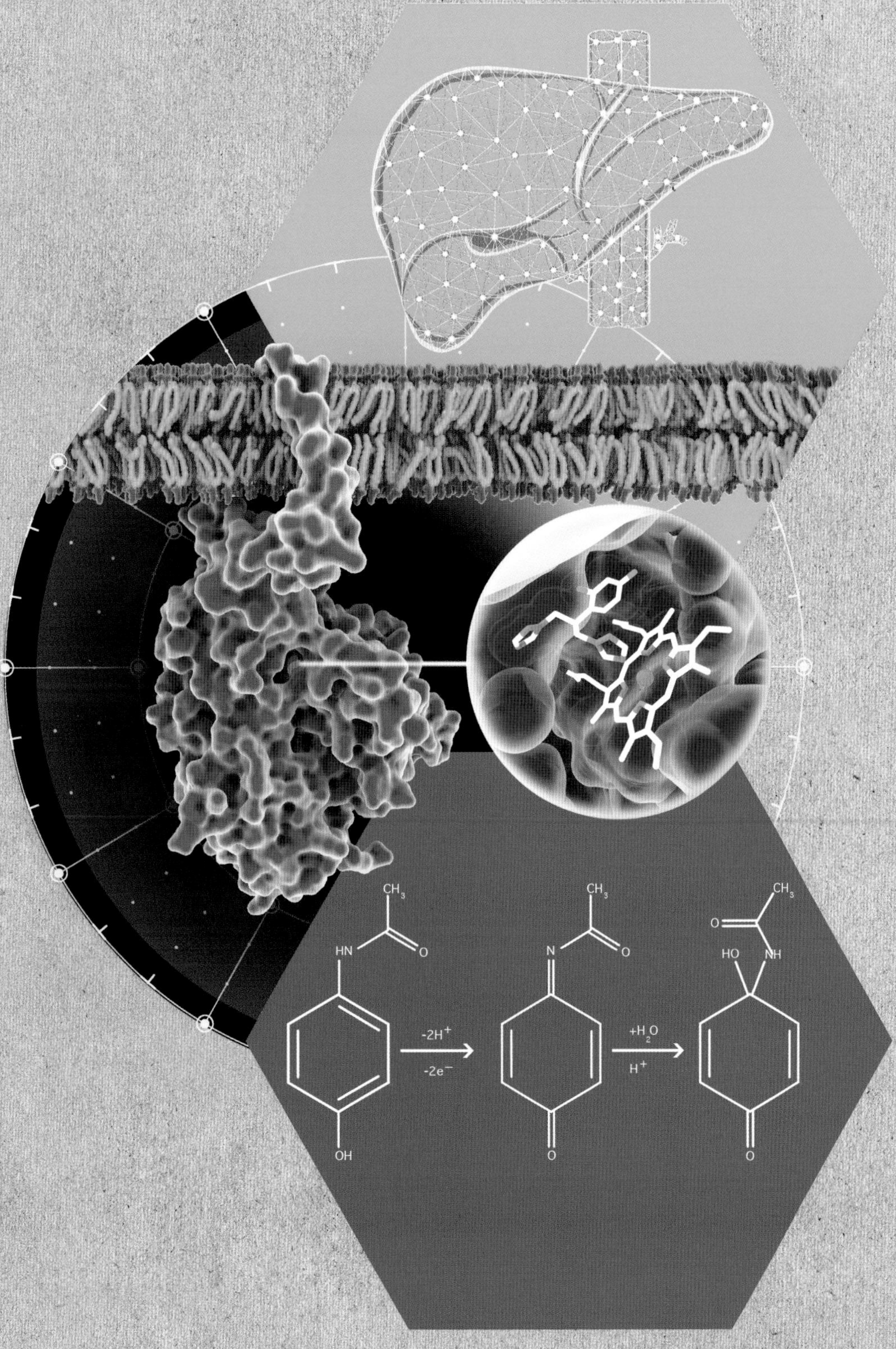

CH3
HN
O
OH
-2H+
-2e−
N
+H2O
H+
HO
NH

ELECTROLYTE BALANCE

the 30-second structure

Balancing life's ions: the kidneys call the shots.

3-MINUTE SYSTEM

In addition to controlling electrolyte balance, the kidneys are a multi-functional command centre that, in conjunction with the heart, lungs, liver, brain and other organs, regulates blood pressure, oxygen concentration, nitrogen waste removal and toxin removal. As part of this effort, the kidneys filter the entire volume of blood into the nephrons every hour.

Electrolytes are ions (charged molecules) such as sodium, potassium and bicarbonate ions that play key roles in many biochemical processes, including muscle contraction, nerve function and the maintenance of fluid pH (acidity). Keeping the appropriate balance among electrolytes in the body is the job of the kidneys, which regulate fluid and electrolyte movement such that any excess electrolytes are eliminated into the urine. They do this using filtration units called nephrons, which are tubes composed of a single layer of cells whose membranes are filled with proteins called ion transporters. These transporters form pores that allow a specific electrolyte to move across the membrane. Fluid and electrolytes are filtered from the bloodstream into the nephron, and electrolytes are excreted through the ion transporters into either the urine or back into the blood according to the body's needs. For example, bicarbonate ion plays a principal role in ensuring the body fluids remain in a narrow pH range. Ordinarily, the kidneys use one type of bicarbonate ion transporter to return filtered bicarbonate ions back into the bloodstream. However, when the pH is too high (such as after taking antacids) a different bicarbonate ion transporter allows for the excess bicarbonate ions to be excreted into the urine, helping to lower the blood pH back to normal.

RELATED TOPICS

See also
LIFE'S GIVE & TAKE
page 24

MEMBRANES, MESSENGERS
& RECEPTORS
page 66

3-SECOND BIOGRAPHIES
SIR WILLIAM BOWMAN
1816–92
English surgeon and anatomist who discovered kidney filtration based on key anatomical observations

HOMER W. SMITH
1895–1962
American physiologist who described several important components of kidney dynamics, including water and electrolyte excretion

30-SECOND TEXT
Steve Julio

The kidneys carry out a whole host of regulatory functions, including maintaining the balance of electrolytes.

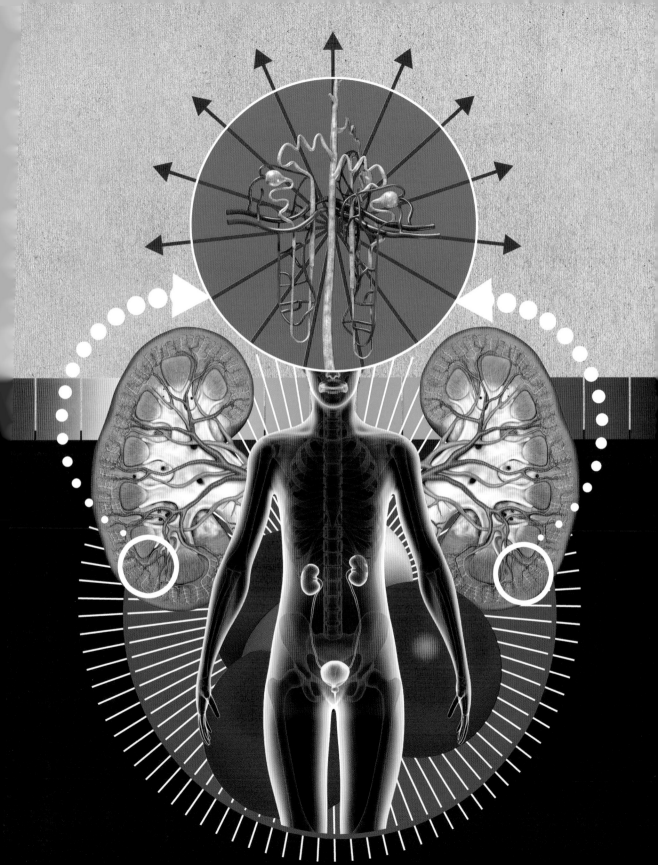

COORDINATING THE DEFENCE

the 30-second structure

When a foreign invader, such as a virus or bacterium, enters our blood, large proteins called antibodies are produced. Antibodies, also called immunoglobulins, recognize and bind the invaders, also known as antigens. In binding, they prevent pathogens from entering cells and target them for destruction. Antibodies contain two longer or heavy protein chains and two shorter or light chains held together by bonds between sulphur atoms, giving a Y-shaped structure with two arms connected to a central stem by a flexible chain. This flexibility permits the antibody to grip the invader at the ends of its arms, where variable amino acids enable the antibody to stick selectively to the invader. The stem of the Y is less variable in structure and contains a carbohydrate chain that is used to communicate with other parts of the immune system. Your immune system adapts and has a memory of past invaders. After initial exposure to an antigen, a second exposure at a later date will result in rapid and more abundant production of the antigen-specific antibodies. Vaccines contain weakened or inactive parts of the antigen that trigger an immune response within the body. If the body encounters the real antigen at a later date, it efficiently produces the antibody that is needed to defeat the invader.

3-SECOND NUCLEUS
Antibodies are large, Y-shaped proteins that are used by our immune system to destroy foreign invaders and keep us healthy.

3-MINUTE SYSTEM
Autoimmune diseases result when the immune system produces a response against one or more of the body's normal constituents and deems it harmful. The immune system can no longer distinguish 'self' and 'non-self' and produces antibodies referred to as autoantibodies that target these otherwise healthy cells, tissues or organs, causing inflammation and damage. Examples of autoimmune diseases include rheumatoid arthritis, multiple sclerosis and lupus.

RELATED TOPICS
See also
CHAINS: BIOLOGICAL POLYMERS
page 38

SWEETNESS: CARBOHYDRATES
page 42

WORKFORCE: AMINO ACIDS & PEPTIDES
page 48

3-SECOND BIOGRAPHIES

EMIL VON BEHRING
1854–1917
German physiologist who discovered the diphtheria antibody

RODNEY PORTER & GERALD EDELMAN
1917–85 & 1929–2014
British biochemist and American biologist who elucidated the chemical structure of an antibody

30-SECOND TEXT
Kristi Lazar Cantrell

Y-shaped proteins, antibodies bind with and neutralize foreign substances such as bacteria and viruses.

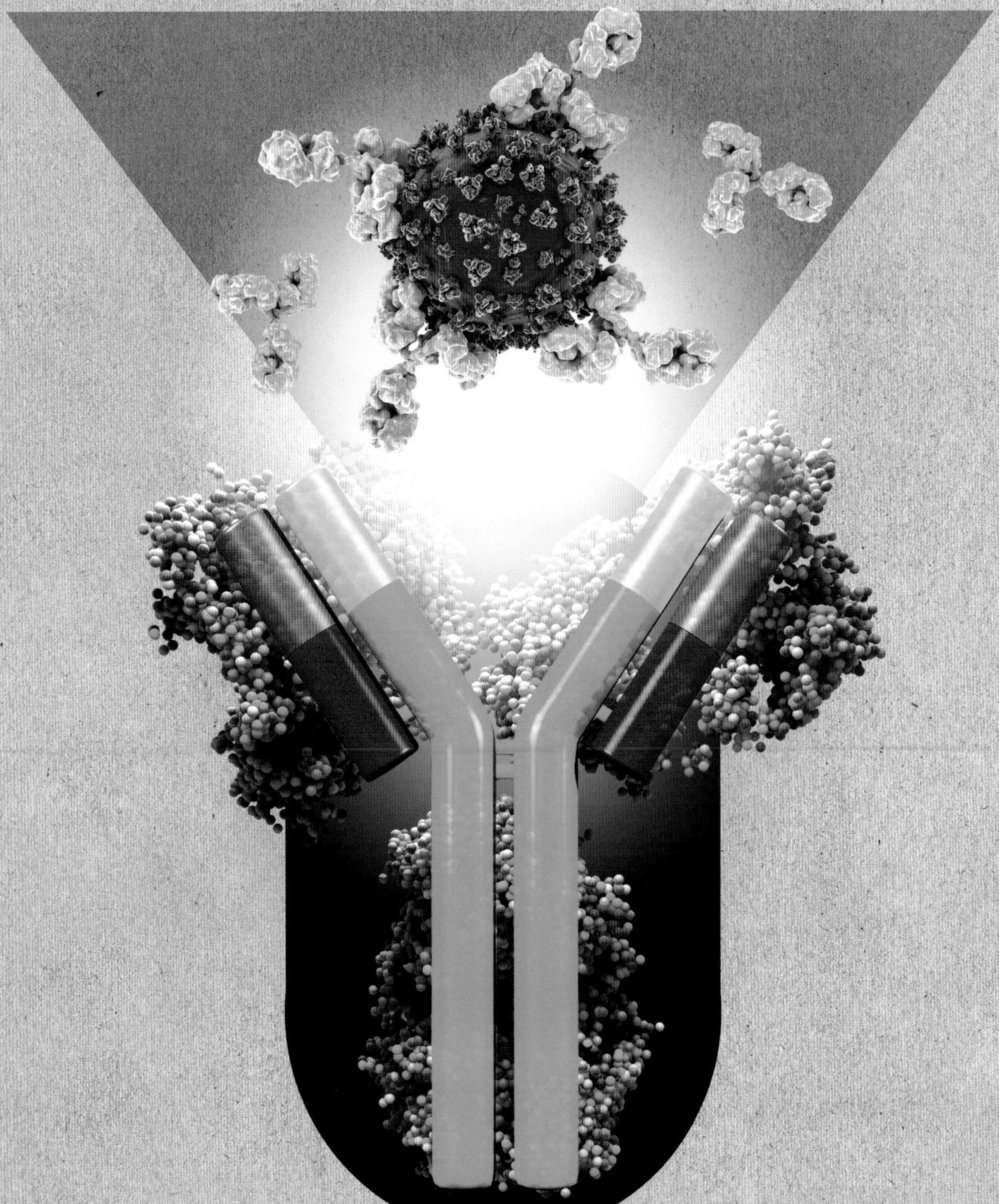

IN SICKNESS & HEALTH

acidosis Condition in which the blood becomes more acidic than normal. If severe enough, acidosis will trigger a cascade of processes that will make the blood even more acidic and, if left untreated, result in death.

adipose tissue Another term for fat tissue, which consists mainly of energy-rich triglycerides.

aerobic Metabolic states and processes that rely on oxygen.

adenosine triphosphate (ATP) A nucleotide that serves as an energy currency, being formed in processes that release energy and broken down in processes that use energy.

anaerobic Metabolic states and processes that do not rely on oxygen.

Cas An acronym for CRISPR-associated proteins, bacterial enzymes that selectively cut CRISPR DNA sequences. Individual Cas proteins are designated by number (for example, Cas9).

cleave The act of breaking apart a biological molecule into its components, usually by adding water to the molecule.

corpus luteum The part of the ovarian follicle left over after an egg is released; it secretes progesterone and other hormones.

CRISPR An acronym for clustered regularly interspaced short palindromic repeats, the type of DNA sequences recognized by Cas proteins in bacteria.

endonuclease An enzyme that cleaves nucleic acids in the middle of a chain.

gene A part of an organism's DNA that encodes for functional RNA or protein.

gene editing The process of modifying an organism's DNA by inserting, deleting or changing DNA sequences.

human chorionic gonadotropin Hormone secreted by cells surrounding a developing embryo that prevents menstruation by stimulating the corpus luteum to keep secreting progesterone.

ketone body Any of three small molecules (acetoacetate, beta-hydroxybutyrate and acetone) produced as a metabolic fuel for tissues during fasting, intense exercise and in some metabolic disorders, notably diabetes mellitus.

luteinizing hormone A hormone secreted by the pituitary gland that triggers ovulation – egg release from the dominant ovarian follicle recruited in a menstrual cycle.

oncogene A gene that, if altered, might result in cancer. Usually, such genes are involved in the regulation of cell division, cell death or DNA repair, which if damaged increases the susceptibility of the cell to uncontrolled proliferation.

ovarian follicle One of many sacs found in the ovaries, each of which contains a single immature egg (oocyte) and hundreds of supporting cells.

phosphocreatine The phosphate-bearing form of creatine, which in the muscles transfers its phosphate to replenish ATP depleted during the first few seconds of exercise.

pituitary gland A small round pea-sized gland at the base of the brain. It secretes hormones into the blood that regulate many body systems, including the function of other hormone-secreting glands. Because of this the pituitary gland is considered the 'master regulator' of the endocrine system.

purine base A class of nucleobase comprising adenine (A) and guanine (G).

pyrimidine base A class of nucleobase comprising cytosine (C), thymine (T) and uracil (U) that are derivatives of the organic molecule pyrimidine.

retinoblastoma protein (Rb) A tumour-suppressor protein inactive in several cancers that is named after retinoblastoma, a cancer of the retina.

retrovirus A virus filled with RNA that is converted by the enzyme reverse transcriptase to DNA in the host cell.

reverse transcriptase The enzyme that converts RNA to DNA – the reverse of transcription.

steroid hormone A type of lipid-based hormone secreted by the adrenal cortex, testes (in males) and ovaries in the placenta (in females), that is involved in the regulation of many bodily functions. Steroid hormones involved in the regulation of reproduction are referred to as sex hormones.

THINGS WE NEED

the 30-second structure

3-SECOND NUCLEUS
Food delivers the necessary components for three basic functions: provision of energy, provision of raw materials for building our cells and regulation of metabolism.

3-MINUTE SYSTEM
Humans cannot synthesize two fatty acids (both polyunsaturated, meaning they contain multiple carbon-carbon double bonds) and nine of the 20 amino acids of proteins. We characterize these nutrients as essential and need to obtain them in adequate amounts from the diet. Also, we cannot make most of the vitamins, with the notable exception of vitamin D, which we produce in the skin with the sunlight's energy.

With the exception of the oxygen we breathe in, all we need to live comes from the food we eat or drink. Nutrition relies heavily on biochemistry for deciphering the composition of foodstuffs and understanding their function in the body at the level of cells and molecules. Of the hundreds of food components, a few dozen enjoy the status of nutrient, that is, they serve one or more of three basic functions: (a) provision of energy, (b) provision of raw materials for building our cells and (c) regulation of metabolism. We recognize five classes of nutrients: carbohydrates (serving function a), fats (serving all three functions), proteins (also serving all three), vitamins (serving function c) and minerals (serving functions b and c). Adequate, but not excessive, intake of all nutrients forms the basis for health, performance in everyday activities and well-being. Carbohydrates (both complex, such as starch, and simple, such as sugar) usually provide most of our daily energy, followed by fats, which are mainly composed of triglycerides. Dietary proteins contribute little to our energy needs and mainly provide amino acids for building our own proteins. Thirteen vitamins (nine water-soluble and four fat-soluble) regulate numerous metabolic processes. Finally, 15 minerals (ten metal and five non-metal) participate in building our tissues and regulating metabolism.

RELATED TOPICS
See also
LIPIDS
page 40

SWEETNESS: CARBOHYDRATES
page 42

WORKFORCE: AMINO ACIDS & PEPTIDES
page 48

3-SECOND BIOGRAPHIES
KAZIMIERZ (CASIMIR) FUNK
1884–1967
Polish biochemist who, in 1912, introduced the concept of vitamins

GEORGE BURR
1896–1990
American biochemist who, in 1929, introduced the concept of essential fatty acids

30-SECOND TEXT
Vassilis Mougios

The body converts essential nutrients into the chemical compounds it needs for energy and optimum health.

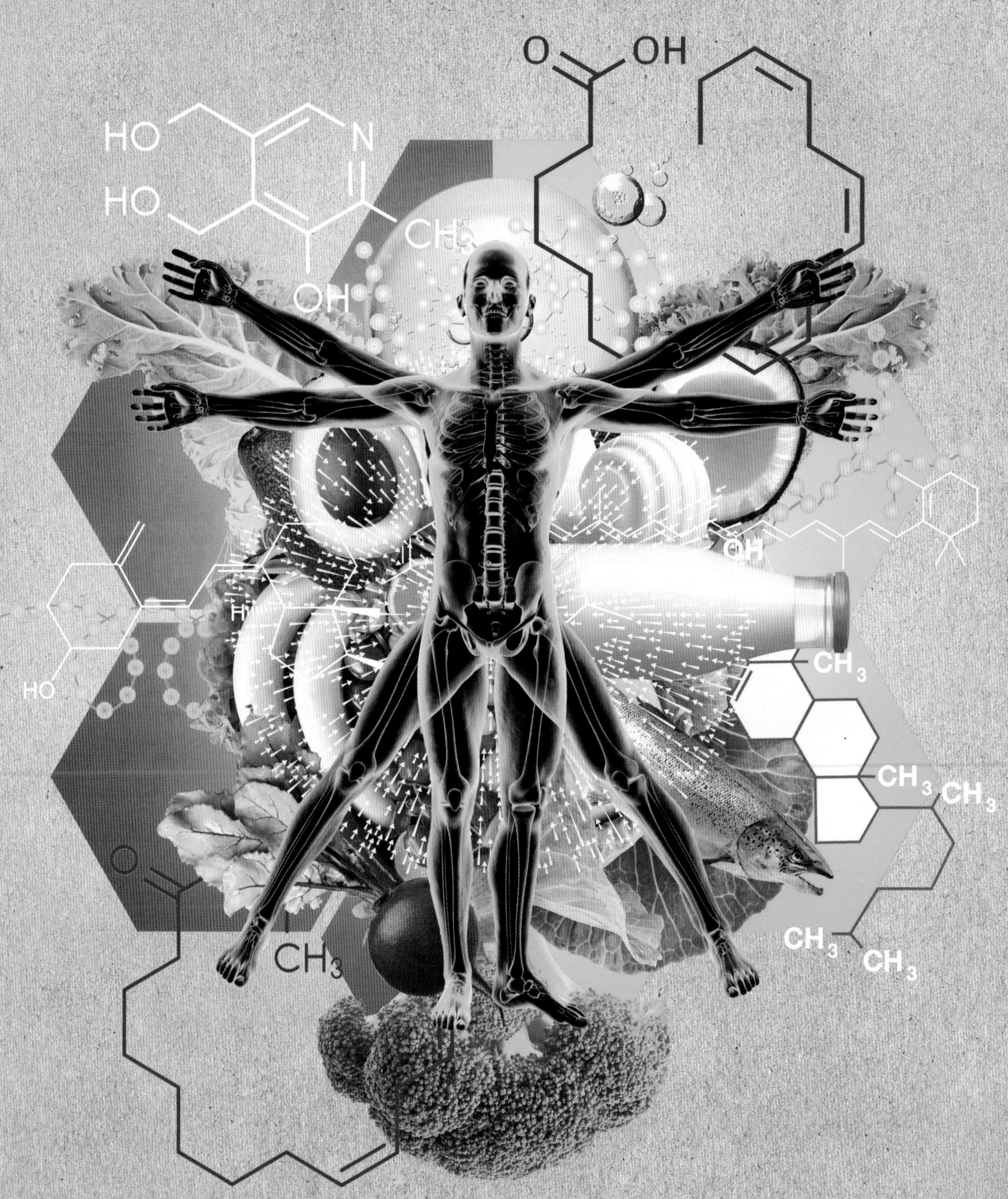

HO
HO
N
CH3
OH
O
OH
OH
HO
CH3
CH3
CH3
CH3
CH3
O
CH3

STAYING FIT

the 30-second structure

Physical exercise imposes great demands on the body, primarily in the form of increased energy for muscle activity. ATP is the direct energy source, but we have only enough for three seconds of maximal activity in reserve. Luckily, a number of indirect chemical energy sources are available that replenish ATP. In muscle the molecule phosphocreatine replenishes ATP very rapidly (in just one reaction), but its amount is limited. Carbohydrates (mainly in the form of muscle and liver glycogen) yield much more ATP, in two ways: anaerobic (producing fast energy) and aerobic (producing more sustained energy). Lipids (mainly in the form of adipose tissue and muscle triglycerides) yield the most ATP, but at slow rates. In contrast, protein breakdown provides minimal energy during exercise; rather, resistance training promotes protein synthesis in muscle, resulting in muscle growth. Numerous molecular mechanisms regulate the relative contribution of each energy source and the mix used depends on the exercise parameters, characteristics of the exerciser and environmental factors. All in all, exercise alters metabolism and, by doing so, galvanizes the body to be more resilient. When repeated regularly, exercise promotes sport performance, fitness, good health and protection against chronic diseases. These benefits are seen in all ages.

3-SECOND NUCLEUS
Exercise alters metabolism and, by doing so, lays the foundations for sport performance, fitness and good health, from childhood to old age.

3-MINUTE SYSTEM
Exercise rarely relies on a single energy source. Rather, it draws energy from a mixture of sources that form an energy continuum, from fast to slow (phosphocreatine, carbs anaerobically, carbs aerobically, lipids) and from little to effectively unlimited (in the same order). This is why it is misleading to characterize exercise as aerobic or anaerobic: in reality, there is an aerobic and an anaerobic component in almost all exercises.

RELATED TOPICS
See also
USING BIG CARBS
page 76

USING SMALL CARBS
page 78

LONG-TERM STORAGE
& BURNING FATS
page 80

MOVING
page 120

3-SECOND BIOGRAPHIES
PER-OLOF ÅSTRAND
1922–2015
Swedish physiologist who made fundamental contributions to many fields of exercise science

BENGT SALTIN
1935–2014
Swedish physiologist who carried out profound research in muscle physiology and biochemistry

30-SECOND TEXT
Vassilis Mougios

Biochemical reactions resulting from regular exercise help keep the body fit and resilient.

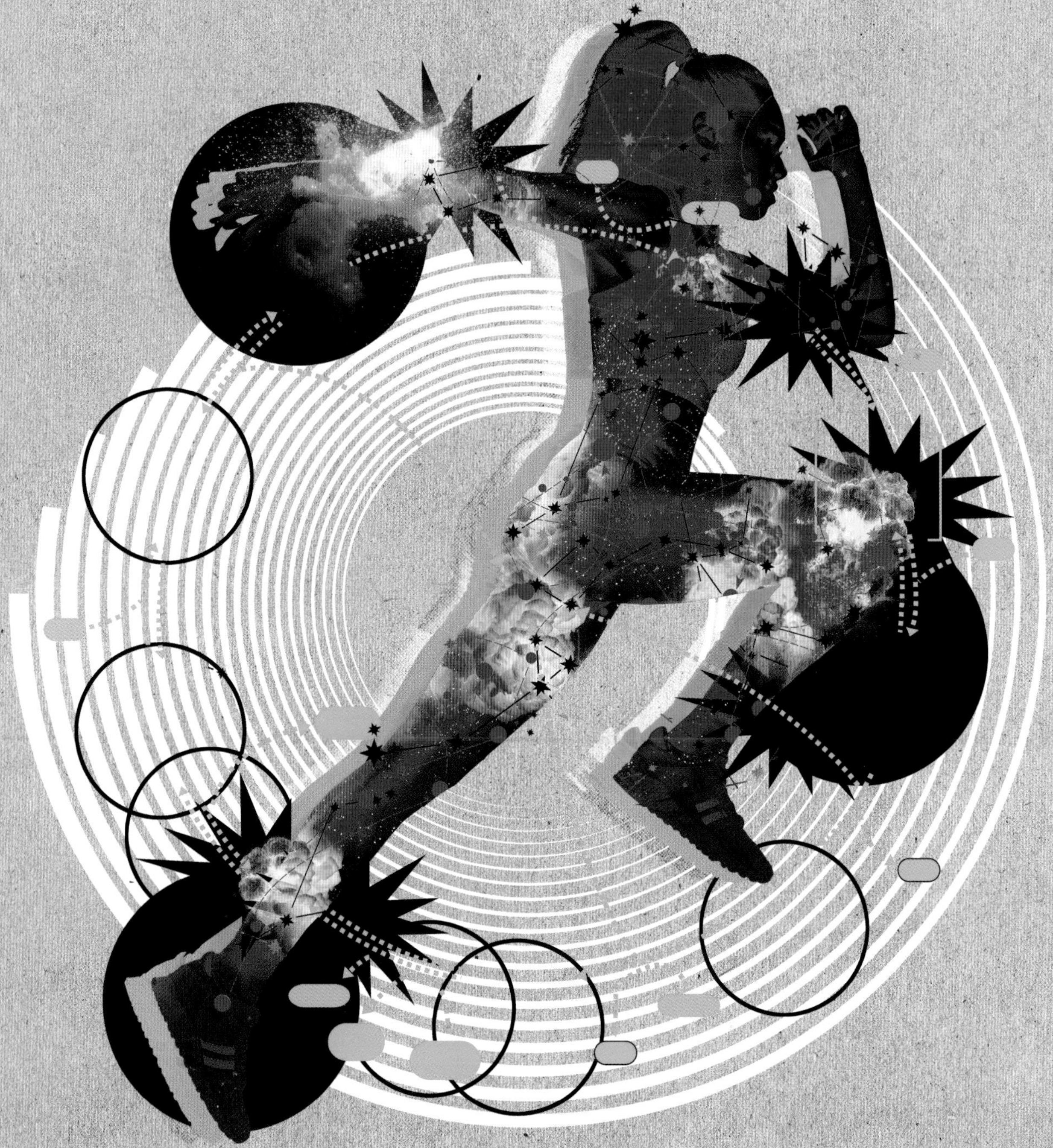

PROGENY & THE PILL

the 30-second structure

The changes a woman's body undergoes to prepare for pregnancy are triggered by hormones. As the menstrual cycle begins, the pituitary gland secretes follicle-stimulating hormone, a small protein that binds to receptors on the surface of follicles in the ovaries, where it triggers a cascade of cellular processes that cause the follicles to grow and develop. The follicles in turn secrete the steroid hormone estradiol, which works on other cells by diffusing through their cell membrane to reach the interior. There, the estradiol binds to a protein that carries it into the cell nucleus and changes the make-up of proteins that the cell produces. Through this process, estradiol secreted by the dominant follicle suppresses the growth of other follicles and stimulates the pituitary to secrete luteinizing hormone, a protein hormone that triggers the dominant follicle to release its egg and causes it to transform into the corpus luteum, which secretes more hormones. This includes progesterone, which, along with estradiol, prepares the uterus for pregnancy by causing its lining to thicken. If the egg is fertilized, the embryo secretes human chorionic gonadotropin, a peptide hormone that stops the cycle and keeps the corpus luteum producing progesterone until the placenta takes over. If not, the corpus luteum decays, progesterone levels drop and menstruation occurs.

3-SECOND NUCLEUS
The menstrual cycle is regulated by peptide hormones secreted by the pituitary glands and steroid hormones produced in the ovary; hormones secreted by the embryo and placenta hold it in suspension during pregnancy.

3-MINUTE SYSTEM
The active ingredients in the first birth control pill, Enovid, were 5 mg norethynodrel and 75 mcg mestranol, two hormones that manipulate the menstrual cycle to prevent pregnancy. The mestranol is a synthetic estradiol that suppresses ovulation while the norethynodrel is a synthetic progesterone that tricks the brain and pituitary gland into acting as if pregnancy had occurred.

RELATED TOPICS
See also
MEMBRANES, MESSENGERS & RECEPTORS
page 66

READING THE BOOK
page 106

3-SECOND BIOGRAPHIES
GREGORY PINCUS
1903–67
American zoologist who helped develop the first oral contraceptive after finding that progesterone inhibits ovulation

GEORGE ROSENKRANZ, CARL DJERASSI & LUIS ERNESTO MIRAMONTES
1916–2019, 1923–2015 & 1925–2004
Mexican (Miramontes and Rosenkranz) and Bulgarian-American (Djerassi) chemists who synthesized the first effective artificial contraceptive

30-SECOND TEXT
Stephen Contakes

Hormones prepare the body for pregnancy; contraceptive pills manipulate this process.

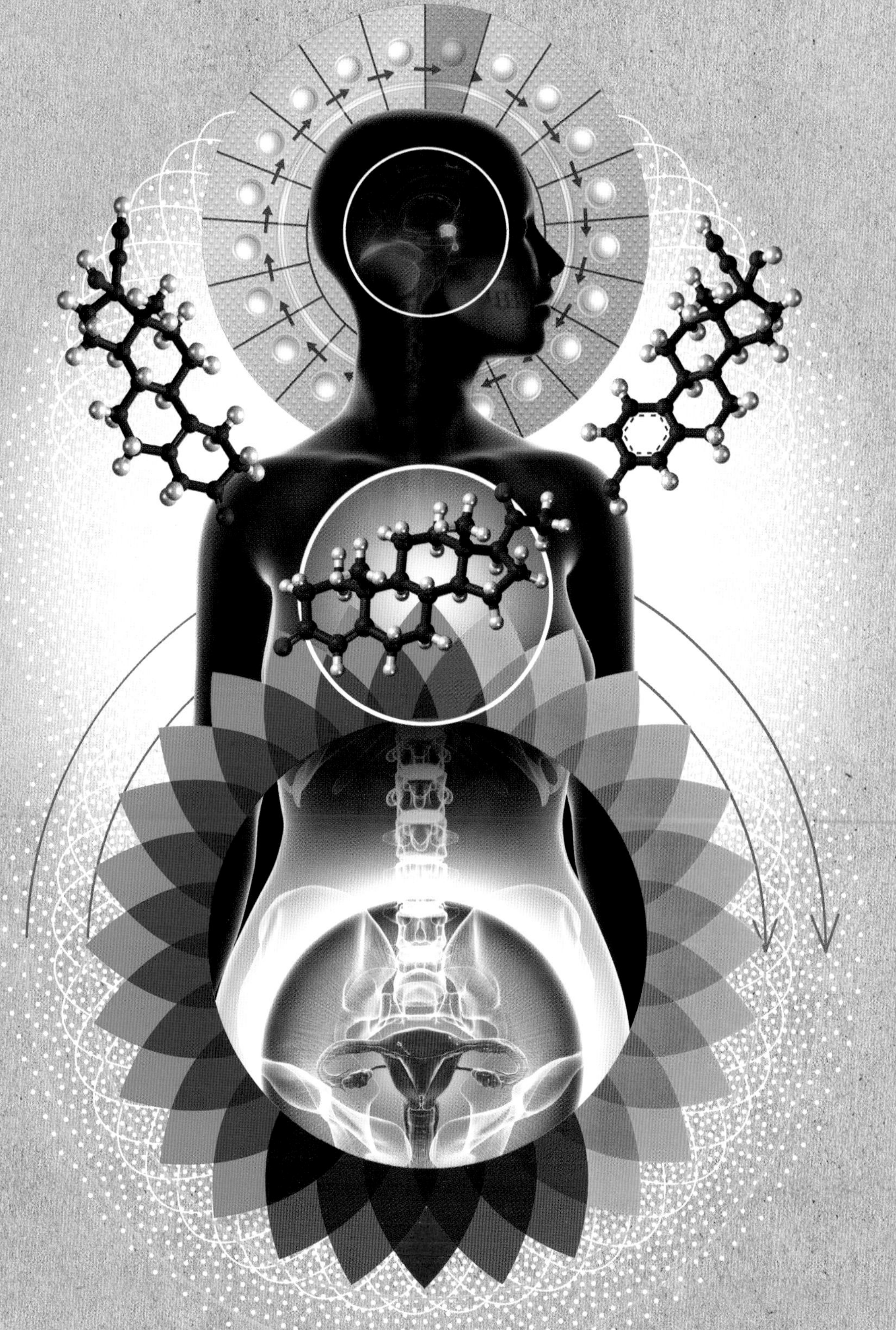

MAINTAINING THE RIGHT BALANCE

the 30-second structure

The hallmark of the metabolic disease diabetes mellitus is elevated levels of the common dietary sugar glucose in the bloodstream. Normally, the pancreas releases the hormone insulin immediately after a meal, which promotes the uptake of newly-digested glucose into the liver, thereby stabilizing blood glucose concentrations. In the diabetic, who either does not produce or cannot respond to insulin, blood glucose levels remain high, which induces a cascade of abnormal responses that can lead to a fatal diabetic coma. Without insulin, the liver begins to release acidic ketone bodies, an alternate cellular fuel source, lowering blood pH. The elevated blood glucose overwhelms the kidney, which fails to regulate water balance, and the resulting decrease in blood fluid volume lowers blood pressure. Inefficient circulation compromises the distribution of oxygen, rendering cells incapable of normal metabolism, and they switch to an alternate form that produces an acidic metabolic by-product, exacerbating the lowered pH of the blood to levels that compromise organ function. This metabolic acidosis, along with the simultaneous drop in blood pressure, is an acute medical emergency. As long as a diabetic can administer insulin appropriately after a meal, these adverse effects are completely avoidable.

3-SECOND NUCLEUS
Diabetes is a blood glucose disease that can cause the body to spiral into a metabolic crisis.

3-MINUTE SYSTEM
Perhaps very few physicians today would agree to diagnose diabetes using the preferred method from several hundred years ago: tasting the patient's urine. Elevated blood glucose increases the amount filtered into the kidney, which results in abnormally high amounts of glucose excreted into the urine. The sweet taste of a diabetic's urine would cause physicians to prescribe ineffective diet modifications and bloodletting. Not until the twentieth-century discovery of insulin did successful treatment become possible.

RELATED TOPICS
See also
USING SMALL CARBS
page 78

MOBILIZING THE CELL
page 88

BLOOD, BREATH & BALANCE
page 118

ELECTROLYTE BALANCE
page 128

3-SECOND BIOGRAPHIES
EDWARD ALBERT
SHARPEY-SCHAFER
1850–1935
English physiologist who first proposed that diabetes was associated with insulin

FREDERICK BANTING
& CHARLES BEST
1891–1941 & 1899–1978
Canadian physician and American physiologist who were the first to isolate insulin and use it to treat diabetes

30-SECOND TEXT
Steve Julio

Insulin and glucagon, secreted by the pancreas, regulate blood glucose levels.

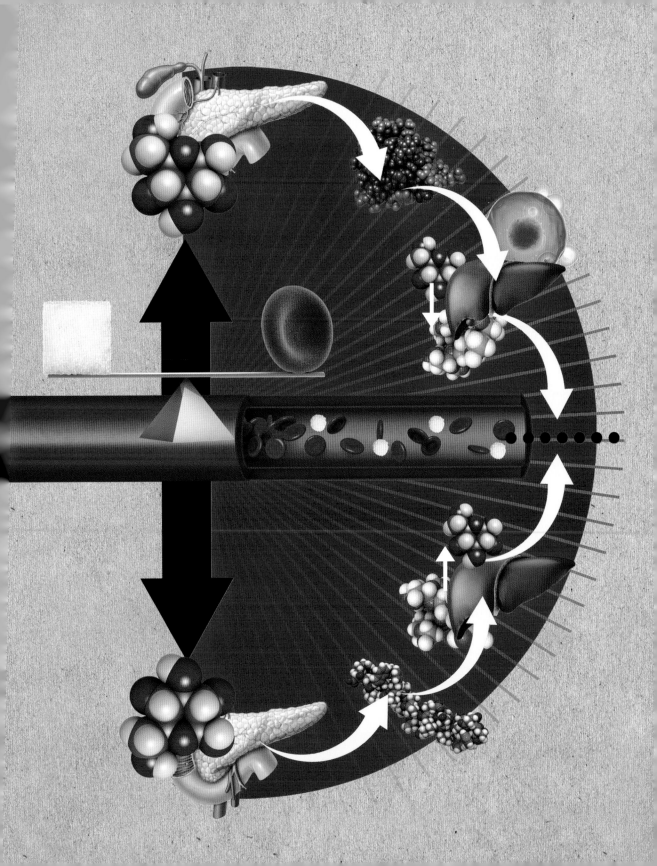

REPELLING THE HIJACKERS

the 30-second structure

Viruses are single-minded machines – their sole purpose is to make more viruses. Consisting of coiled-up nucleic acid surrounding a protein coat called a capsid, a virus binds to its target cell, becomes internalized, hijacks the cell's machinery to replicate itself and is released to infect more cells. This replication process disrupts cell function or even destroys cells, causing illness. For respiratory viruses such as influenza (flu), damage to cells in the lungs can impair oxygen uptake, leading to respiratory distress. Vaccines can protect us but are not always effective – influenza vaccines are a best guess based on circulating strains, and there is still no vaccine against HIV. Antiviral drugs, while not cures, can help. Antiviral drugs are highly specific. For influenza, a familiar drug is oseltamivir (marketed as Tamiflu®), which can reduce the severity and duration of infection. Two proteins found on the surface of influenza are hemagglutinin and neuramidase (the 'H' and 'N' in flu strains such as H1N1). Following infection, the new influenza copies are initially stuck to the cells with viral hemagglutinin bound to cell surface receptor proteins – these receptors need to be cleaved by neuramidase to liberate the virus. Oseltamivir blocks this cleavage, causing the virus to remain stuck, preventing subsequent infection.

3-SECOND NUCLEUS
Viruses are lean, mean, infecting machines, and sometimes our bodies need a little chemical help to fight them off: enter the antivirals.

3-MINUTE SYSTEM
Human immunodeficiency virus (HIV) antivirals are given as a multidrug cocktail to ensure efficacy should the virus develop resistance to any one component, a concern given how fast HIV mutates. Indeed, this is why early HIV treatment with a single drug (AZT, for example) ultimately failed. A typical three-drug cocktail targeting important proteins for HIV infection comprises an integrase inhibitor, which blocks insertion of virus genes into human DNA, and two different nucleoside reverse transcriptase inhibitors, which block replication of viral genes.

RELATED TOPICS
See also
SHAPE: PROTEIN STRUCTURE
page 56

COORDINATING THE DEFENCE
page 130

GERTRUDE ELION & GEORGE HITCHINGS
page 146

3-SECOND BIOGRAPHIES
RICHARD SHOPE
1901–66
American virologist who discovered the human influenza virus during the 1930s, years after the 1918 flu epidemic

MARTINUS BEIJERINCK & DMITRI IVANOVSKY
1851–1931 & 1864–1920
Dutch microbiologist and Russian botanist who (independently) are credited with describing the first virus – tobacco mosaic virus

30-SECOND TEXT
Andrew K. Udit

Viruses enter via the cell membrane surface; some antivirals bind to cell receptor proteins to block this action.

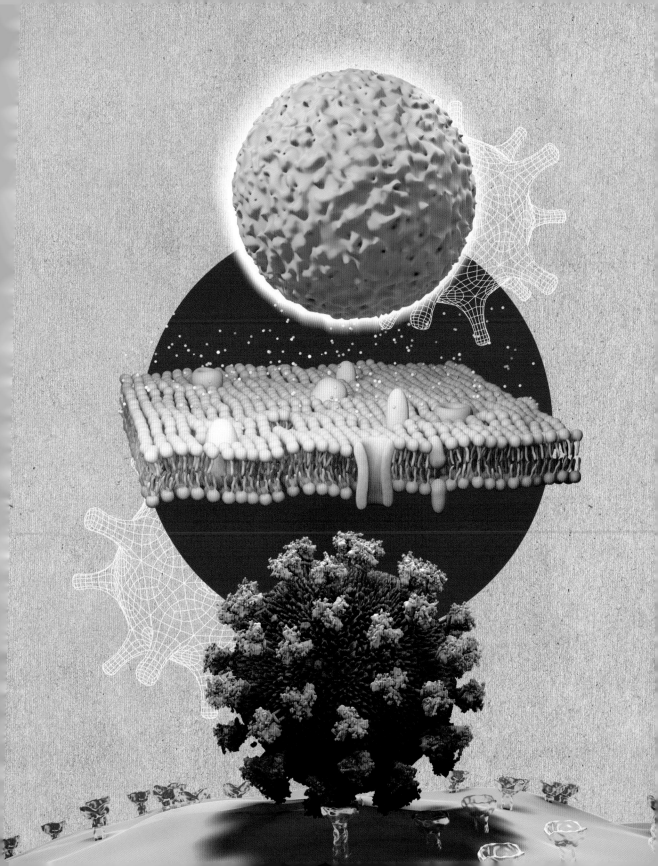

18 April 1905
George Herbert Hitchings born in Hoquiam, Washington, USA

23 February 1918
Gertrude Elion born in Manhattan, New York, USA

1933
Hitchings receives a PhD from Harvard

1937
Elion receives a bachelor's degree in chemistry from Hunter College

1942
Hitchings joins the Biochemistry Department at Wellcome Research Laboratories in Tuckahoe, New York

1944
Elion joins George Hitchings' lab at Wellcome Research Laboratories

1947
Elion and Hitchings find 2,6-diaminopurine works as chemotherapy

1967
Elion becomes Head of Biochemistry at Wellcome Research Laboratories. Hitchings becomes Vice President in Charge of Research of Burroughs Wellcome

1968
Elion designs the antiviral diaminopurine arabinoside, which inhibits multiple viruses

1971
Hitchings becomes President of the Burroughs Wellcome Fund

1977
Elion discovers that acyclovir, related to diaminopurines, inhibits herpesviruses infections

1988
Elion and Hitchings awarded the Nobel Prize in Physiology or Medicine

27 February 1998
Hitchings dies in Chapel Hill, North Carolina, USA

21 February 1999
Elion dies in Chapel Hill, North Carolina, USA

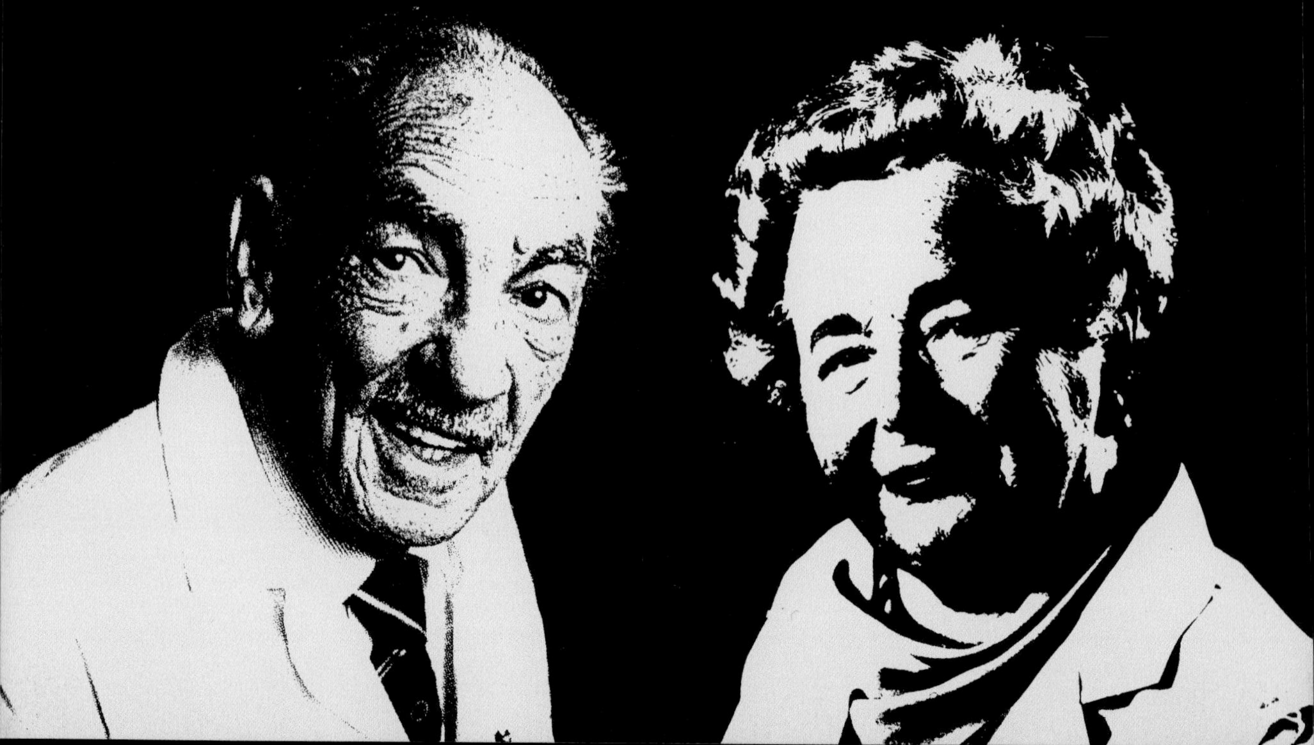

GERTRUDE ELION & GEORGE HITCHINGS

Gertrude Elion and George Hitchings used organic chemistry and biochemical logic to make new antiviral drugs. They shared the Nobel Prize in Physiology or Medicine for this in 1988 with Sir James W. Black. Hitchings and Elion earned bachelor's degrees in chemistry in 1927 and 1937, respectively. Hitchings received a PhD from Harvard in 1933 during the Great Depression, and encountered what he described as 'a nine-year period of impermanence, both financial and intellectual'. Elion also endured instability, made worse by sexism: 'Jobs were scarce and the few positions that existed in laboratories were not available to women.'

In 1942 Hitchings started his own laboratory at the Wellcome Research Laboratories, and two years later Elion joined as an organic chemist. The laboratory combined multiple disciplines, from chemistry to immunology to virology, all working to develop new drugs. Hitchings' strategy was to use defective versions of molecules that bacteria needed to block them from making nucleotides, which are the building blocks of DNA and RNA in all life, including viruses.

In what seemed like a dead end at the time, Hitchings and Elion found that one purine compound, 2,6-diaminopurine, showed activity both as an antiviral and chemotherapeutic (anti-cancer) agent, but it was too toxic. Over the next 20 years, Elion developed expertise in building and testing other purine-related molecules, including thioguanine, azathioprine and allopurinol. During the 1950s, biochemists learned much more about the pathways by which purines are synthesized, used and degraded in life. This information would help turn 2,6-diaminopurine into a potent drug.

When Hitchings took an administrative post at Burroughs Wellcome Co. in 1967, Elion became head of the laboratory. She noticed a report that the purine adenine arabinoside inhibited viruses and reasoned that 2,6-diaminopurine arabinoside might do so as well. Experiments showed she was right, and suggested other chemical modifications of purines to try. Elion tried these and made acyclovir, which was very potent and selective against several viruses, and non-toxic to mammalian cells. Acyclovir has been used since 1980 to treat herpes and shingles infections. Elion's reasoning also led to other purine-based drugs against cancer (nelarabine), leukaemia (mercaptopurine) and gout (allopurinol).

Because these discoveries were based on rational modification of biochemical structures using organic chemistry, they opened the way to other rationally designed drugs. Elion and Hitchings therefore not only found effective drugs but also founded a method of thinking more targeted, and more successful, than previous trial-and-error drug discovery.

Ben McFarland

WHEN CELLS BREAK BAD

the 30-second structure

One of the most important features the cells in your body possess is the ability to reproduce in a controlled way. You developed from a single-cell embryo that repeatedly divided and, under the influence of hormonal signals, split off into more specialized cells in carefully defined directions. The cells that constitute your tissues and organs developed as the hormones shaped which genes they translated into protein, whether they divided and whether they died. Eventually, you stopped developing. Cells came into close contact with one another and stopped dividing, and the rate at which old cells died and new cells formed became approximately equal under the influence of the regulatory mechanisms that control cell development and death. When the parts of a cell's DNA that control this process (oncogenes) are damaged beyond repair, a tumour, or mass of abnormally growing cells, forms. For instance, normally the retinoblastoma or Rb protein keeps cells from replicating its DNA. However, if the gene that encodes Rb is damaged, the cell does not make the Rb it needs to stop dividing. Depending on exactly how DNA is damaged in a tumour cell, the cell might also gain the ability to divide in the presence of other cells and move throughout the body. When this occurs the result is cancer, a malignant tumour.

3-SECOND NUCLEUS
Cancer is a disease in which a cell's DNA becomes corrupted in ways that lead it to divide uncontrollably.

3-MINUTE SYSTEM
A variety of factors shape whether a cell experiences the sort of DNA damage needed to give rise to a malignant tumour. High-energy radiation such as X-rays and UV light or carcinogenic chemicals directly modify DNA, while inherited defects in oncogenes give cells a head start on the way to forming a tumour. For example, mutations in the BRCA1 and BRCA2 genes increase the risk of breast and ovarian cancer by making it easier for your cells to lose the ability to repair defective DNA.

RELATED TOPICS
See also
LIFE'S LETTERS: NUCLEOTIDES
page 46

MAINTAINING THE CODE
page 102

3-SECOND BIOGRAPHIES

THEODOR BOVERI
1862–1915
German zoologist who first proposed that cancer is caused by defects in chromosomes

ERIC BOYLAND
1905–2002
English chemist who first hypothesized that some cancers result when the liver converts seemingly inert chemicals to cancer-causing metabolic intermediates

J. MICHAEL BISHOP & HAROLD E. VARMUS
1936– & 1939–
American virologists who discovered how retroviruses cause cancer

30-SECOND TEXT
Stephen Contakes

Some types of DNA damage result in abnormal cells that divide uncontrollably.

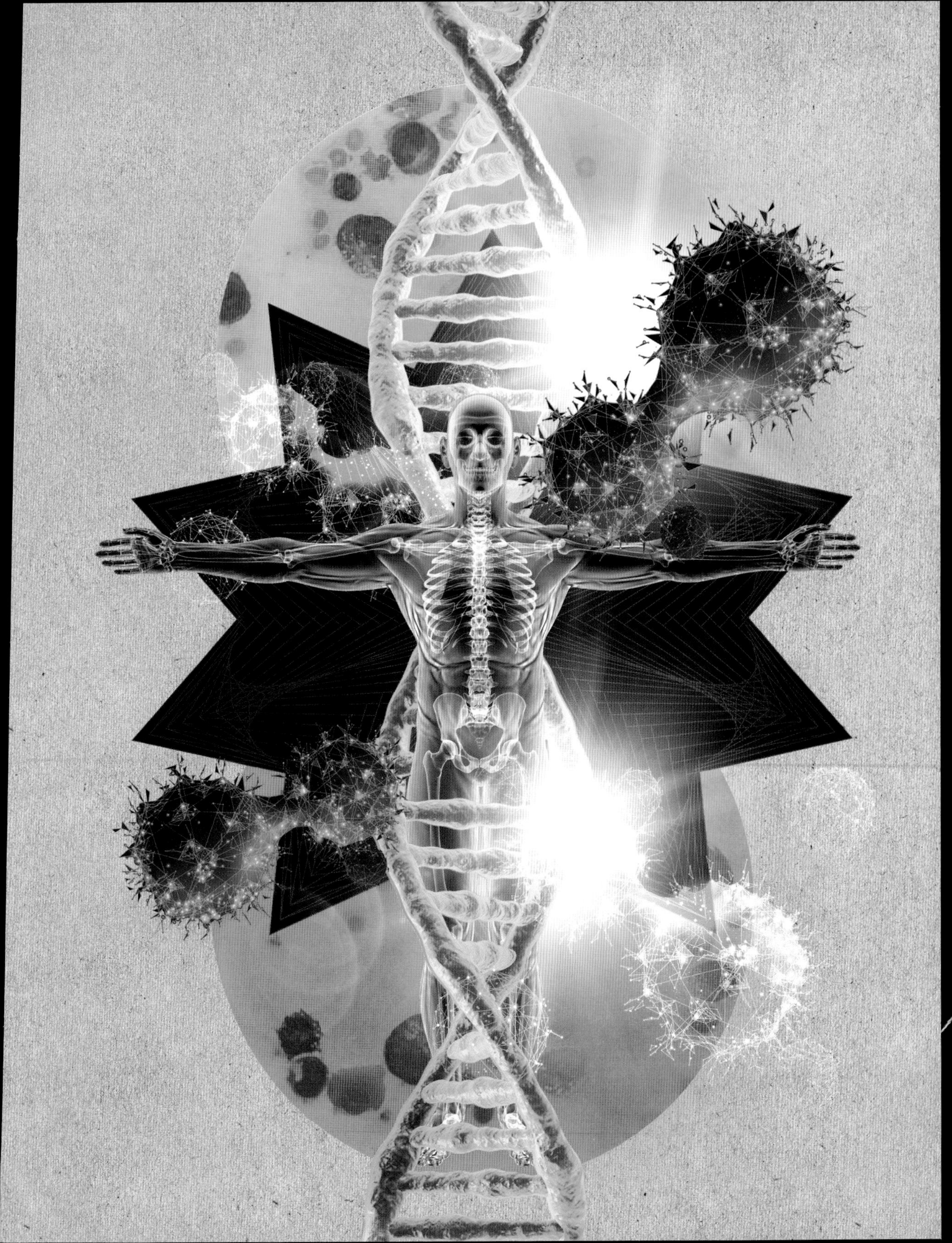

GENE EDITING

the 30-second structure

Just as old films **were edited**
by cutting out parts and pasting them back
together, the genetic code may be edited by
cutting and stitching DNA back together. Such
DNA editing uses enzymes called endonucleases
to cut or modify DNA at precisely defined sites.
For example, in CRISPR gene editing a CRISPR-
associated endonuclease such as Cas9 cuts
DNA at a location determined by the sequence
of a piece of 'guide RNA'. The DNA is then
modified as the cell uses enzymes such as DNA
polymerase to repair the cut. When cells'
DNA repair mechanisms are left to their own
devices, they tend to add or remove bases and
so inactivate any genes present. DNA can be
edited more precisely using a specially designed
piece of DNA called a repair template. The repair
template contains the edits to be made, along
with bases that can pair with those near the cut.
When the repair template pairs with the DNA at
the cut it directs the repair process to add to or
edit the DNA as it repairs the cut. Another form
of gene editing uses a broken endonuclease
(that cannot cut DNA) to find a specific DNA
sequence and an attached enzyme to change
the nearby code by chemically altering one
nucleobase into another.

RELATED TOPICS
See also
LIFE'S LETTERS:
NUCLEOTIDES
page 46

MAKING LIFE GO: ENZYMES
page 68

MAINTAINING THE CODE
page 102

REPELLING THE HIJACKERS
page 144

3-SECOND BIOGRAPHIES
PAUL BERG
1926–
American biochemist who
developed methods for
inserting new DNA into an
existing DNA sequence

JENNIFER DOUDNA &
EMMANUELLE CHARPENTIER
1964– & 1968–
American and French
biochemists who developed
the CRISPR/Cas9 gene-editing
system

30-SECOND TEXT
Stephen Contakes

*Gene editing changes
organisms' DNA; it may
one day be used to treat
genetic diseases.*

3-SECOND NUCLEUS
Technologically and
medically useful plants,
animals and bacteria can
be made by modifying
organisms' DNA using
enzymes that selectively
cut one or more of
its strands.

3-MINUTE SYSTEM
Genetic engineering has
been used to make vitamin
A-enriched rice, farmed
salmon that grow more
rapidly and bacteria that
produce human insulin for
the treatment of diabetes.
In medicine relatively
precise and straightforward
technologies such as
CRISPR/Cas gene editing
are actively being
investigated as a way to
make human immune cells
that fight cancer. One day
they might be used to
correct debilitating
genetic defects or, more
controversially, design
babies with particular
characteristics.

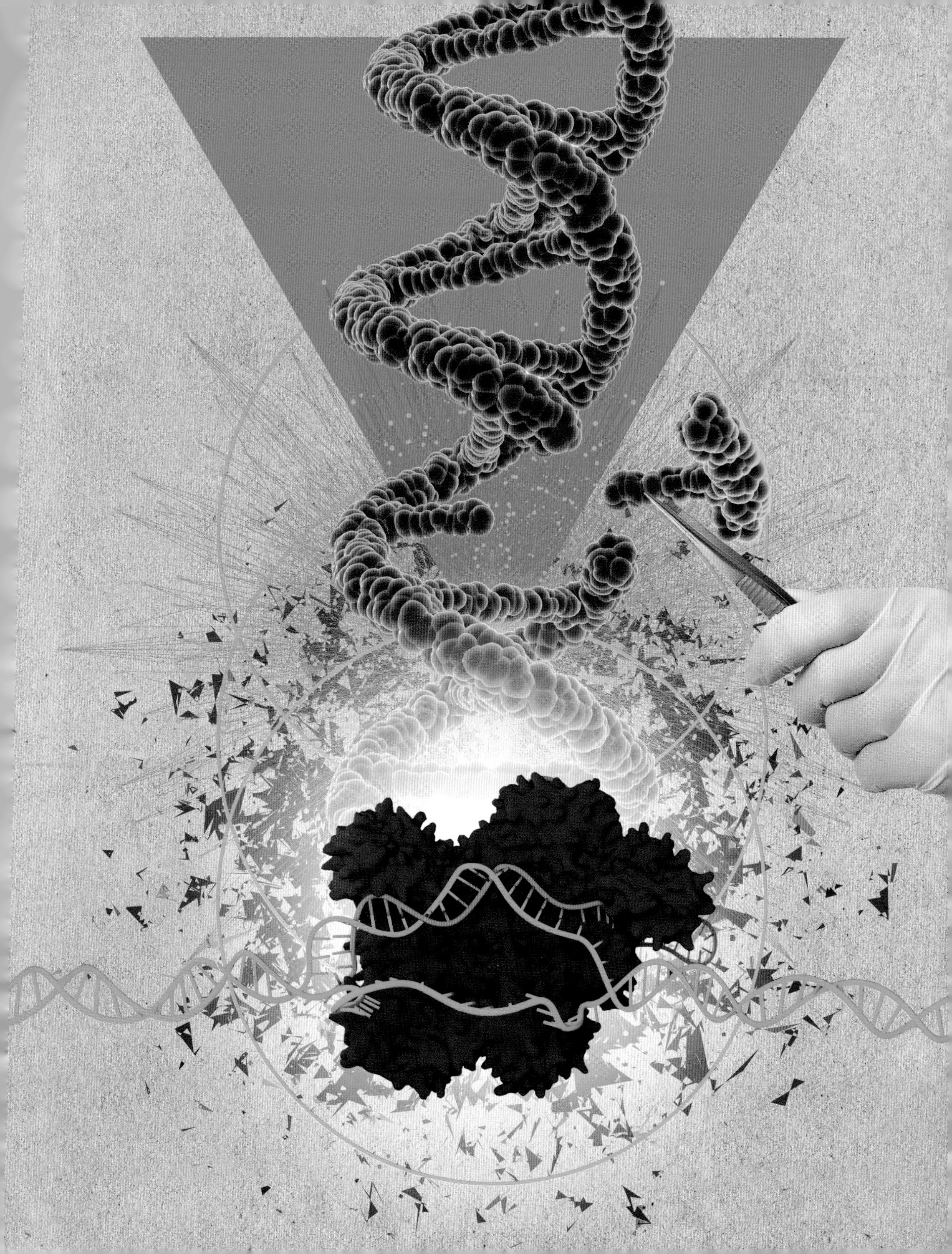

APPENDICES

RESOURCES

BOOKS

Biochemistry
Jeremy M. Berg, Lubert Stryer,
John L. Tymoczko & Gregory J. Gatto
(Macmillan, 9th edition, 2019)

Biochemistry: The Molecular Basis of Life
James R. McKee & Trudy McKee
(Oxford University Press, 2020)

Biochemistry: The Molecules of Life
Richard Bowater, Laura Bowater
& Tom Husband
(Oxford University Press, 2020)

Color Atlas of Biochemistry
Jan Koolman & Karl-Heinz Roehm
(Thieme, 3rd edition, 2012)

*A Crack in Creation: Gene Editing and the
Unthinkable Power to Control Evolution*
Jennifer A. Doudna & Samuel H. Sternberg
(Houghton Mifflin, 2017)

*The Eighth Day of Creation: Makers
of the Revolution in Biology*
Horace Freeland Judson
(Cold Spring Harbor Laboratory Press,
Commemorative edition, 1996)

Exercise Biochemistry
Vassilis Mougios
(Human Kinetics, 2nd edition, 2019)

*Fundamentals of Biochemistry:
Life at the Molecular Level*
Donald Voet, Judith G. Voet
& Charlotte W. Pratt
(Wiley, 5th edition, 2016)

Harper's Illustrated Biochemistry
Victor W. Rodwell et al.
(McGraw-Hill, 31st edition, 2018)

Lehninger Principles of Biochemistry
David L. Nelson & Michael M. Cox
(Freeman, 7th edition, 2017)

The Machinery of Life
David S. Goodsell
(Springer, 2nd edition, 2009)

*The Making of the Fittest: DNA and the
Ultimate Forensic Record of Evolution*
Sean B. Carroll
(W. W. Norton & Company, 2006)

The Manga Guide to Biochemistry
Masaharu Takemura, Kikuyaro & Office Sawa
(No Starch Press, 2011)

Nature's Building Blocks: An A–Z Guide to the Elements
John Emsley
(Oxford University Press, 2011)

On Food and Cooking: The Science and Lore of the Kitchen
Harold McGee
(Simon & Schuster, 2007)

Oxygen: The Molecule that Made the World
Nick Lane
(Oxford University Press, 2016)

What is Life?: How Chemistry Becomes Biology
Addy Pross
(Oxford University Press, 2016)

A World From Dust: How the Periodic Table Shaped Life
Ben McFarland
(Oxford University Press, 2016)

WEBSITES

The Biochemical Society biochemistry booklets page
https://biochemistry.org/education/schools-and-fe-colleges/biochemistry-booklets-for-a-level/

The Biology Libretexts biochemistry page
https://bio.libretexts.org/Bookshelves/Biochemistry

The National Center for Biotechnology Information (USA)
https://www.ncbi.nlm.nih.gov/

The Protein Data Bank Learning Page
http://pdb101.rcsb.org/

NOTES ON CONTRIBUTORS

EDITOR

Stephen Contakes is Associate Professor of Chemistry at Westmont College in Santa Barbara, California, where he currently teaches courses in inorganic, analytical and physical chemistry. He completed BSc degrees from Lehigh University, a PhD from the University of Illinois at Urbana-Champaign and postdoctoral work at the California Institute of Technology. With a background in organometallic, bioinorganic and biophysical chemistry, his research interests involve the preparation of molecular assemblies and photoactive nanoparticles with a view towards environmental applications.

CONTRIBUTORS

Steve Julio received his PhD at the University of California, Santa Barbara, and is currently Professor of Biology at Westmont College in Santa Barbara. He teaches courses in molecular biology, physiology and infectious diseases. His research interests are in bacterial pathogenesis, where he studies the bacterial disease mechanisms that allow *Bordetella*, the causative agent of whooping cough, to establish and maintain an infection in the mammalian respiratory tract.

Kristi Lazar Cantrell is Associate Professor of Chemistry at Westmont College in Santa Barbara, California, where she teaches courses in general and organic chemistry, and biochemistry. She has been honoured as Westmont's outstanding teacher of the year. Her research involves protein aggregation, including alpha-helical and beta-sheet fibril assembly, with some projects related to neurodegenerative diseases.

Ben McFarland is Professor of Biochemistry at Seattle Pacific University in Washington. Dr McFarland and his students design and synthesize proteins that are part of the immune response, and then measure how those proteins bind using physical chemistry. These have included antibodies, natural killer cell receptors and proteins that trap bacterial siderophore molecules. Dr McFarland is interested in how chemistry relates to the big questions in biology and geology, such as how biological evolution coincided with chemical and mineral evolution. He has stocked a mini-museum outside his office with his collection of the elements and minerals important to life.

Vassilis Mougios is Professor of Exercise Biochemistry at the Aristotle University of Thessaloniki (AUTh), Greece, with a BSc in chemistry from the University of Athens, Greece, and a PhD in biochemistry from the University of Illinois. He has taught exercise biochemistry, exercise physiology, sport nutrition and ergogenic aspects of sport to undergraduates and graduates in his home country and abroad. He has conducted research on muscle activity, exercise metabolism, biochemical assessment of athletes and sport nutrition and has authored or co-authored more than 100 papers in international scientific journals. He is the author of *Exercise Biochemistry* (Human Kinetics, 2019).

Dr Anatoli Petridou is part of the laboratory teaching staff at Aristotle University of Thessaloniki (AUTh), Greece. She holds a BSc in Physical Education and Sport Science from AUTh and a BSc in biochemistry and biotechnology from University of Thessaly, Greece. She received both her MSc in human performance and health and her PhD in exercise biochemistry from AUTh. She has also received a postdoctoral fellowship from the AUTh Research Committee. Her research interests include the effects of physical exercise and nutrition on metabolism and gene expression. Dr Petridou is author or co-author of 38 peer-reviewed publications, with more than 1,000 citations.

Andrew K. Udit received his HBSc from the University of Toronto and his PhD from the California Institute of Technology and completed postdoctoral studies at the Scripps Research Institute. He is currently an Associate Professor of Chemistry at Occidental College in Los Angeles, California, where he teaches courses in general chemistry, biochemistry and writing classes in the college's core programme. His research in chemical biology includes biocatalysis with P450 enzymes and developing virus-like particles to act as heparin antagonists.

INDEX

ACKNOWLEDGEMENTS

The publisher would like to thank the following for permission to reproduce copyright material:

Alamy/Photo 12: 104r

GSK Heritage Archives: 146r

Mark Liberman: 104c

Shutterstock/Emmily: 15; Kateryna Kon: 15; artemide: 15; sciencepics: 15; Nerthuz: 15; SciePro: 15; Anton Nalivayko: 17; John MacNeill: 17; SciePro: 17 Jubal Harshaw: 17; Pixxsa: 17; andamanec: 19; Nixx Photography: 19; Madrock 24: 19; 3d_man: 19; Le Do: 19; StudioMolekuul: 21; yaruna: 21; Stu Shaw: 21; Susse_n: 21; ifong: 21; Pixxsa: 23; sciencepics: 23; Azat Valeev: 23; Sergey Bogomyako: 23; gritsalak karalak: 25; Korosi Francois-Zoltan: 25; zita: 27; Sanjaya viraj bandara: 27; Designua: 27; chemistrygod: 27; Quality Stock Arts: 35; NPeter: 35; bobbyramone: 35; John MacNeill: 39; SciePro: 39; MIKHAIL GRACHIKOB: 39; StudioMolekuul: 39; dny3d: 39; Explode: 41; Lightspring: 41; StudioMolekuul: 41; Alila Medical Media: 41; StudioMolekuul: 43; Steve Cukrov: 43; Nickolay Grigroriev: 43; StudioMolekuul: 47; IrenD: 47; Mad Dog: 47; Pixxsa: 47; Designua: 49; KRAHOVNET: 49; StudioMolekuul: 49; StudioMolekuul: 55; ibreakstock: 55; StudioMolekuul: 61; Kateryna Kon: 63; StudioMolekuul: 65; Pixxsa: 67; StudioMolekuul: 67; J. Marini: 67; ibreakstock: 67; molaruso: 69; StudioMolekuul: 75; Pixxsa: 75; Nixx Photography: 75; Nerthuz: 75; Liu zishan: 75; StudioMolekuul: 81; Design_Cells: 81; SciePro: 81; maximmmmum: 81; Funny Drew: 81; StudioMolekuul: 83; Pixxsa: 89; Legolena: 89; Anton Nalivayko: 89; Kateryna Kon: 89; StudioMolekuul: 89; Maryna Olyak: 89; SciePro: 89; StudioMolekuul: 95; Nixx Photography: 95; Chyrko Olena: 95; Aldona Griskeviciene: 95; 3d_man: 95; KRAHOVNET: 95; StudioMolekuul: 97; roshan_pics: 99; Leszek Czerwonka: 99; YANGCHAO: 99; phochi: 99; pro500: 99; Quality Stock Arts: 101; Pixxsa: 101; StudioMolekuul: 101; vonzur: 103; bestber: 103; StudioMolekuul: 103; KRAHOVNET: 103; MohamadOuaidat: 107; Meletios Verras: 107; Jose Luis Calvo: 109; eranicle: 109; a Sk: 109; Monika7: 109; joker1991: 111; Vandathai: 111; Chaikom: 111; Damsea: 111; orinocoArt: 111; sirtravelalot: 111; Igor Petrushenko: 117; ibreakstock: 117; StudioMolekuul: 117; mybox: 119; Designua: 119; crystal light: 119; StudioMolekuul: 119; Magcom: 119; Monika7: 119; Pixxsa: 119; martan: 121; sciencepics: 121; Antti Heikkinen: 121; Jose Luis Calvo: 121; Coksawatdikorn: 121; nobeastsofierce: 121; Designua: 125; Login: 125; whitehoune: 125; Aromaan: 125; Johan Swanepoel: 125; Pixxsa: 129; StudioMolekuul: 129; Magic mine: 129; 131: Christoph Burgstedt; Soleil Nordic: 131; Design_Cells: 131; Login: 131; tuulijumala: 131; ustas7777777: 131; Alexander Raths: 137; Beautyimage: 137; Nik Merkulov: 137; JIANG HONGYAN: 137; Valentina Razumova: 137; Diana Taliun: 137; Hortimages: 137; Picsfive: 137; OxfordSquare: 137; Bogmenko Evgenia: 139; Master1305: 139; Djem: 139; Michael Zduniak: 139; Fireofheart: 139; mStudioVector: 139; Borbely Edit: 139; Social Media Hub: 141; Marochkina Anastasilia: 141; Kateryna Kon: 141; gomolach: 141; Tefi: 141; StudioMolekuul: 143; Andrea Danti: 143; Kateryna Kon: 143; eranicle: 143; Maryna Olyak: 143; mayakova: 143; CG_dmitriy: 145; Design_Cells: 145; jivacore: 145; Explode: 149; Dim Dimich: 149; vonzur: 149; fusebulb: 149; Warunyaporn Promphat: 149; Jozsef Bagota: 149; Vikpit: 149; andriano.cz: 151; Explode: 151; art_of_sun: 151; Meletios Verras: 151; BAIVECTOR: 151

Smithsonian: 25

Wikimedia Commons: 25, 26, 35, 37, 44, 58, 61, 63, 77, 79, 81, 83, 84, 87, 97, 99, 104l, 107, 109, 111, 122, 125, 127, 137, 141, 146l

All reasonable efforts have been made to trace copyright holders and to obtain their permission for the use of copyright material. The publisher apologizes for any errors or omissions in the list above and will gratefully incorporate any corrections in future reprints if notified.